PLASTICS ADDITIVES

PLASTICS ADDITIVES

AN INDUSTRIAL GUIDE

Third Edition

Volume I

by

Ernest W. Flick

NOYES PUBLICATIONS
WILLIAM ANDREW PUBLISHING, LLC
Norwich, New York, U.S.A.

Library of Congress Catalog Card Number: 00-108117
ISBN: 0-8155-1464-6
Printed in the United States

Published in the United States of America by
Noyes Publications / William Andrew Publishing, LLC
Norwich, New York, U.S.A.

10 9 8 7 6 5 4 3 2 1

Library of Congress Cataloging-in-Publication Data

Flick, Ernest W.
        Plastics additives: an industrial guide / by Ernest W. Flick. --
    3rd ed.
              p.   cm.
        Includes index.
        ISBN 0-8155-1464-6
        1. Plastics--Additives. I. Title.
TP1142.F58   2001
668.4'11--dc20
                                                    00-108117
                                                    CIP

To
Kaitlin Leeann Margarite Brown

Preface

The Third Edition of this useful book is divided into three volumes. Volume I describes almost 1,000 plastics additives which are currently available to industry. It is the result of information received from 76 industrial companies and other organizations. The data represents selections from manufacturers' descriptions made at no cost to, nor influence from, the makers or distributors of these materials. Only the most recent information has been included. It is believed that all of the products listed here are currently available, which will be of utmost interest to readers concerned with product discontinuances.

Plastics additives, a complex and growing group of minerals and chemical derivatives, account for 15 to 20% by weight of the total volume of plastic products marketed. They are produced by a large number of companies for a wide variety of customers and needs. Growth of the use of these additives, which impart one or more desirable properties to resins, is relatively strong, about 3.5 to 4% per year. Environmental constraints, however, have imposed rigorous performance requirements on many products, thus placing added expenses on the development costs of these materials.

Plastics additives are divided into logical sections in three volumes.

Volume I:

    I. Adhesion Promoters

    II. Anti-Fogging Agents

The Table of Contents of each volume is organized to also serve as a subject index. Each raw material is located in the section which is most applicable. The reader seeking a specific raw material should check each section which could possibly apply. In addition to the above, two further sections are included in each volume: a list of Suppliers' Addresses and a Trade Name Index. These will be extremely helpful to the reader.

The following information is supplied for each product, as available, in the manufacturer's own words:

> Company name and product category.
>
> Trade names and product numbers.
>
> Product description: a description of the product's main features, as described by the supplier.

The three volumes of this book will be of value to technical and managerial personnel involved in the preparation of products made with these plastics additives as well as to the suppliers of the basic raw materials.

My fullest appreciation is expressed to the companies and organizations who supplied the data included in this book.

November, 2000   Ernest W. Flick

NOTICE

Contents and Subject Index

SECTION III: ANTIOXIDANTS

SECTION IV: ANTI-STATIC AGENTS

SECTION V: ANTIBACTERIALS/FUNGICIDES/
MILDEWCIDES

SECTION VI: BONDING, BLOWING
AND FOAMING AGENTS

SECTION VII: DISPERSANTS

SECTION VIII: FILLERS AND EXTENDERS

SECTION IX: FIRE AND FLAME RETARDANTS/SMOKE SUPPRESSANTS

Section I
Adhesion Promoters

ADMTronics Unlimited, Inc.: Primers/Adhesives:

AQUAFORTE 108-W:
Water-Based Primer/Adhesive for Extrusion and Laminations

An excellent extrusion, lamination and printing primer for use on a wide variety of accepting substrates. Along with its excellent grease and oil resistance, Aquaforte 108-W is also water and moisture resistant. High bond strengths are achieved with only minute applicaton levels, approximately one undiluted gallon of Aquaforte 108-W for 20 to 30 reams (60,000 to 90,000 ft2) of substrate.

Aquaforte 108-W should be applied to the substrate and dried. The extradite, lamination or printing inks will then rigorously bond to the primed surface.

Characteristics:
    Odorless
    Non-toxic
    Non-flammable
    Non-corrosive
    Long shelf life

Uses:
    Extrusion Primer
    Lamination
    Printing Primer

Compatible Combinations:
LDPE-Surlyn-Printing Inks to accepting substrates: Polyester, Polypropylene, Nylon, Cellophane, Vinyl, PVDC, Aclar, Aluminum Foil, Polycarbonate, Paper and Board.

POLAQUA W7:
Water-Based Primer/Adhesive for Extrusions and Laminations to Clay Coated Paper Surfaces

Polaqua W7 is an excellent extrusion, lamination and printing primer for use on a wide variety of accepting substrates and in particular clay coated paper surfaces.

Applying Polaqua W7 on a clay coated surface will allow for extremely high bond strengths from heat sealing of LDPE to the surface or for extrusions of LDPE to the clay surface. This makes Polaqua W7 suitable for priming clay coated boards for back sealing improvement in cup and box applications. In addition, Polaqua W7 will improve the bond of the clay to the paper surface.

Polaqua W7 should be applied to the substrate and dried. The extrudate, lamination or printing inks will then rigorously bond to the primed surface. The substrate may be primed and rewound for subsequent use at a later date up to 90 days with no blocking or reduction in bond strengths.

Characteristics:
    Odorless
    Non-toxic
    Non-flammable
    Non-corrosive
    Long shelf life

Uses:
    Extrusion Primer
    Lamination
    Printing Primer

ADMTronics Unlimited, Inc.: Primers/Adhesives (Continued):

Polaqua 73F:
 Water-Based Primer/Adhesive for Extrusions and Laminations

 Polaqua 73F is an excellent extrusion, lamination and printing
primer for use on a wide variety of accepting substrates. Along
with excellent grease and oil resistance, Polaqua 73F is also
water and moisture resistant.
 Polaqua 73F should be applied to the substrate and dried.
The extrudate, lamination or printing inks will then rigorously
bond to the primed surface.

Characteristics: Uses:
 Acrylic odor   Extrusion primer
 Non-toxic   Lamination adhesive
 Non-flammable   Printing ink primer
 Non-corrosive
 Long shelf life

Compatible Combinations:
 LDPE-Surlyn-Printing Inks to accepting substrates:
 Polyester, polypropylene, nylon, cellophane, vinyl, PVDC,
Aclar, aluminum foil, polycarbonate, paper and board.

Polaqua 103-L:
 Water-based Primer/Adhesive for Polypropylene and Other Films

 Polaqua 103-L is an extrusion, lamination and printing primer
for use on many grades of polypropylene and other films. It
exhibits grease, oil and moisture resistant bond strengths with
only minute application levels.
 Polaqua 103-L is applied to the film surface and dried. For
best results, especially with certain grades of polypropylene,
it is recommended to have the accepting surface corona treated.
The extrudate, lamination or printing inks will then rigorously
bond to the primed surface by heat and pressure application.

Characteristics: Uses:
 Non-toxic   Extrusion primer
 Non-flammable   Lamination
 Non-corrosive   Printing primer
 Water-based

Compatible Combinations:
 LDPE-Surlyn-Printing inks to accepting substrates:
 Polypropylene, polyester, nylon, vinyl, cellophane, aluminum
foil, PVDC, Aclar, polycarbonate, paper and board

ADMTronics Unlimited, Inc.: Primers/Adhesives (Continued)

Polaqua 138:
Aqueous Primer

Polaqua 138 is a water-based primer specifically developed for water resistant bonds of polyethylene to aluminum foil by lamination or extrusion. It is applied by conventional application methods (gravure, flexo, rod, roll, etc.) to the accepting substrate. The primer must be thoroughly and absolutely dry before introduction of the extrudate or lamination. Polaqua 138 should be applied as received using light application weights.

Standard extrusion temperatures are suitable for acceptable bonding. In lamination it is necessary to hot-nip at a minimum of 180 degrees F. Excellent bond strengths will be achieved off-machine with full product resistance properties developed after time.

Polaqua 138 is suitable for combinations of LDPE or Surlyn to aluminum foil. Acceptable combinations of EVA to foil and polyester have been achieved.

Polaqua 138 is formulated from ingredients cleared by the FDA for food packaging applications. It is water-based and complies with the strictest EPA or OSHA restrictions.

SANTEL HR-97:
Printing Primer for Films

Santel HR-97 is a water-based primer for all grades of films which is used to improve printing adhesion and lay down properties. Certain films are difficult to print and inks may not have adequate adhesion for packaging applications. Applying Santel HR-97 to the surface of the film, prior to printing, will greatly improve ink lay down on press and final print adhesion values.

Santel HR-97 is used as supplied and can be applied with conventional coating apparatus such as gravure, flexo, roll, blade, rod, etc. Light application weights are typical but optimum weights should be determined by the user. A typical starting point would be the application of 6 pounds wet per ream (3000 square feet). After applying, the surface should be completely dry before printing. It is non-tacky and can be wound for subsequent use.

Santel HR-97 is also an excellent primer which enhances and permits adhesion of PVDC to LDPE and PVC films. It has been tested with other substrates as well and has been found to perform well.

Santel HR-97 is completely water-based and contains no organic solvents or ozone depleting chemicals.

Chartwell International, Inc.: CHARTWELL Adhesion Promoters:
General Properties:

Chartwell:

B-505.1:
pH, 2%: 4.15
Wt % Active: 25.6
Sp. g: 1.23
Metals (1): 2
Organic group: mercapto

B-505.6:
pH, 2%: 6.40
Wt % Active: 31.5
Sp. g: 1.36
Metals: 2
Organic Group: mercapto

B-515.1:
pH, 2%: 4.50
Wt % active: 25.6
Sp. g: 1.23
Metals: 2
Organic group: pri amine

B-515.1W:
pH, 2%: 4.80
Wt % active: 21.5
Sp. g: 1.09
Metals: 2
Organic group: pri amine

E-515.2:
pH, 2%: 4.50
Wt % active: 21.5
Sp. g: 1.22
Metals: 2
Organic group: pri amine

B-515.4:
pH, 2%: 7.20
Wt % active: 39.6
Sp. g: 1.23
Metals: 1
Organic group: pri amine

B-515.4W:
pH: 7.70
Wt % active: 33.4
Sp. g: 1.20
Metals: 1
Organic group: pri amine

B-515.5W:
pH, 2%: 7.60
Wt % Active: 30.3
Sp. g: 1.19
Metals: 1
Organic group: pri amine

B-516.5:
pH, 2%: 6.80
Wt % Active: 41.6
Sp. g: 1.26
Metals: 1
Organic group: pri & sec amine

B-516.5W:
pH, 2%: 7.30
Wt % active: 36.6
Sp. g: 1.20
Metals: 1
Organic group: pri & sec amine

B-523.2W:
pH, 2%: 5.30
Wt % active: 31.2
Sp. g: 1.19
Metals: 1
Organic group: hydroxy/carboxy

B-523.2WX:
pH, 2%: 5.30
Wt % active: 34.5
Sp. g: 1.19
Metals: 1
Organic group: hydroxy/carboxy
(1.5x)

B-523.6W:
pH, 2%: 7.20
Wt % active: 29.7
Sp. g: 1.20
Metals: 2
Organic group: hydroxy/carboxy

C-523.2:
pH, 2%: 3.40
Wt % active: 22.0
Sp. g: 1.15
Metals: 2
Organic group: hydroxy/carboxy

(1) Denotes number of different metals

Chartwell International, Inc.: CHARTWELL Adhesion Promoters: General Properties (Continued):

Chartwell:

B-525:
  pH, 2%: 4.05
  Wt % Active: 22.5
  Sp. g: 1.08
  Metals (1): 1
  Organic group: carboxy

B-525.1:
  pH, 2%: 4.30
  Wt % active: 27.4
  Sp. g: 1.23
  Metals: 2
  Organic group: carboxy

D-525.3:
  pH, 2%: 4.25
  Wt % active: 25.5
  Sp. g: 1.14
  Metals: 2
  Organic group: carboxy

C-531.1:
  pH, 2%: 4.30
  Wt % active: 25.1
  Sp. g: 1.08
  Metals: 2
  Organic group: C-10 carboxy

D-531.1:
  pH, 2%: 4.00
  Wt % active: 25.1
  Sp. g: 1.24
  Metals: 2
  Organic group: C-10 carboxy

B-535.1:
  pH, 2%: 4.05
  Wt % active: 24.2
  Sp. g: 1.06
  Metals: 2
  Organic group: C-14 hydrocarbon

D-535.1:
  pH, 2%: 4.00
  Wt % active: 24.2
  Sp. g: 1.10
  Metals: 2
  Organic group: C-14 hydrocarbon

B-545.1:
  pH, 2%: 4.35
  Wt % active: 25.6
  Sp. g: 1.23
  Metals: 2
  Organic group: methacrylato

B-600:
  pH, 2%: 4.40
  Wt % active: 22.5
  Sp. g: 1.09
  Metals: 2
  Organic group: sulfido

B-600.6:
  pH, 2%: 6.80
  Wt % active: 31.5
  Sp. g: 1.36
  Metals: 2
  Organic group: sulfido

(1) Denotes number of different metals

Chartwell International, Inc.: CHARTWELL Adhesion Promoters:
General Properties for Higher Solid "H" Products:

Chartwell:

B-505.1H:
 pH, 2%: 4.40
 Wt % Active: 33.7
 Sp. g: 1.29
 Metals (1): 2
 Organic group: mercapto

B-515.1H:
 pH, 2%: 4.55
 Wt % Active: 38.5
 Sp. g: 1.31
 Metals: 2
 Organic group: pri amine

B-515.1WH:
 pH, 2%: 4.60
 Wt % Active: 31.5
 Sp. g: 1.22
 Metals: 2
 Organic group: pri amine

E-515.2H:
 pH, 2%: 4.85
 Wt % Active: 33.0
 Sp. g: 1.27
 Metals: 2
 Organic group: pri amine

B-515.4WH:
 pH, 2%: 6.90
 Wt % Active: 44.0
 Sp. g: 1.27
 Metals: 1
 Organic group: pri amine (1.5x)

B-515.5WH:
 pH, 2%: 6.90
 Wt % Active: 41.1
 Sp. g: 1.26
 Metals: 1
 Organic group: pri amine

B-516.5WH:
 pH, 2%: 6.85
 Wt % Active: 42.4
 Sp. g: 1.24
 Metals: 1
 Organic group: pri & sec amine

B-523.6WH:
 pH, 2%: 7.35
 Wt % Active: 40.5
 Sp. g: 1.30
 Metals: 2
 Organic group: hydroxy/carboxy

C-523.2H:
 pH, 2%: 3.40
 Wt % Active: 30.3
 Sp. g: 1.22
 Metals: 2
 Organic group: hydroxy/carboxy

B-525.1H:
 pH, 2%: 4.30
 Wt % Active: 36.9
 Sp. g: 1.29
 Metals: 2
 Organic group: carboxy

C-531.1H:
 pH, 2%: 3.90
 Wt % Active: 38.0
 Sp. g: 1.17
 Metals: 2
 Organic group: C-10 carboxy

B-535.1H:
 pH, 2%: 4.16
 Wt % Active: 32.9
 Sp. g: 1.11
 Metals: 2
 Organic group: C-14 hydrocarbon

C-535.1H:
 pH, 2%: 4.22
 Wt % Active: 33.9
 Sp. g: 1.15
 Metals: 2
 Organic group: C-14 hydrocarbon

All CHARTSIL products also available as "H" grades
(1) Denotes number of different metals

Dow Corning Corp.: DOW CORNING Coupling Agents/Adhesion Promoters:

Dow Corning Z-6020 Silane:
Features: Aminofunctional; improves adhesion of resins to various substrates; hydrophobes bonds for longer service life.
Typical Applications: Epoxies, phenolics, melamines, nylons, PVC, acrylics, polyolefins, polyurethanes, nitrile rubbers and fiberglass-reinforced thermoplastics; used to couple inorganic fillers or reinforcing materials with resins; surface pretreatment of fillers and reinforcers.

Dow Corning Z-6030 Silane:
Features: Methacrylate, improves adhesion of resins to various substrates, hydrophobes bonds for longer service life.
Typical Applications: Free-radical crosslinked polyester rubber, polyolefins, styrenics, acrylics; used to couple inorganic fillers or reinforcing materials with resins; surface pretreatment of fillers and reinforcers.

Dow Corning Z-6032 Silane:
Features: Styrylamine cationic; very versatile over a broad spectrum of materials and applications; improves adhesion of resins to various substrates; hydrophobes bonds for longer service life.
Typical Applications: Most thermoset and thermoplastic resins; adhesion promoter; used to couple inorganic fillers or reinforcing materials with resins; surface pretreatment of fillers and reinforcers.

Dow Corning Z-6040 Silane:
Features: Epoxy, improves adhesion of resins to various substrates; hydrophobes bonds for longer service life.
Typical Applications: Adhesion promoter for epoxies, urethane, acrylic and polysulfide sealants; used to couple inorganic fillers or reinforcing materials with resins; surface pretreatment of fillers and reinforcers.

Sartomer: Adhesion Promoters:

UV Cure Applications:
CN-704:
   CN-704 is an acrylated polyester adhesion promoter that provides excellent adhesion to polyolefins, such as polyethylene and polypropylene. CN-704 is designed for UV/EB cured inks and coatings. Due to its low acid value, CN-704 can be used in formulations containing tertiary amines. Usage levels of 40-60% are recommended.

SR-9008:
   SR-9008 is an alkoxylated difunctional monomer developed to assist adhesion of UV/EB cured inks and coatings to glass surfaces. It is also an excellent choice for OPV applications. SR-9008 features good weatherability and impact strength. Usage levels of 5-10% are recommended.

SR-9012:
   SR-9012 is a trifunctional acrylate monomer that assists adhesion of UV/EB cured inks and coatings to plastics, metal, and glass. Recommended usage levels are 5-10%.

SR-9016:
   SR-9016 is a metallic diacrylate adhesion promoter for UV/EB cure applications. SR-9016 is water soluble up to 20 wt%. This product promotes adhesion to metals, including steel, tin-free steel, brass, and aluminum. Usage levels up to 5% are recommended

CD-9050:
   CD-9050 is a multifunctional acid ester that promotes adhesion to metals, including aluminum, brass, steel, and tin-free steel. CD-9050 is an excellent choice for low shrinkage. Due to its high acid value, this product is not recommended for use in formulations containing tertiary amines. Usage levels of 3% to 12% are recommended.

CD-9051:
   CD-9051 is a trifunctional version of CD-9050. CD-9051 offers the same adhesion promoting properties as CD-9050. However, due to its trifunctionality, CD-9051 provides faster cure response, greater hardness, and good water resistance. As with CD-9050, CD-9051 is not recommended for use in formulations containing tertiary amines. Usage levels of 3% to 12% by weight are recommended.

CD-9052:
   CD-9052 is a trifunctional acid ester adhesion promoter for UV/EB inks and coatings that offers exceptional cure speed. CD-9052 assists adhesion to difficult substrates, such as plastics, metals, and glass. CD-9052 is not recommended for use in formulations containing tertiary amines. Usage levels of 3-12% are recommended.

Sartomer: Adhesion Promoters (Continued):

Peroxide Cure Applications:
SR-633:
    SR-633 is a metallic diacrylate monomer, developed to
assist rubber-to-metal adhesion in peroxide cure applications.
Recommended use levels are 4-10%.

SR-634:
    SR-634 is a methacrylate version of SR-633. SR-634 features
excellent hardness and tear strength. Usage levels of 4-10%
are recommended.

SR-705:
    SR-705 is a metallic diacrylate monomer that assists adhesion
to metals and plastics in peroxide cure applications. SR-705 is
water-soluble and features excellent fast cure response. Usage
levels of 4-10% are recommended.

SR-706:
    SR-706 is a modified metallic diacrylate monomer with similar
adhesion promoting properties to SR-705. However, SR-706 offers
faster cure response. Usage levels of 4-10% are recommended.

SR-708:
    SR-708 is a metallic dimethacrylate monomer developed to
assist adhesion to metals and plastics in peroxide cure applications. SR-708 features hot tear strength. Usage levels of 4-10%
are recommended.

SR-709:
    SR-709 is a metallic monomethacrylate monomer that assists
adhesion to metals and plastics in peroxide cure applications.
SR-709 features good tear strength. Recommended usage levels
are 4-10%.

SR-9009:
    SR-9009 is a trifunctional methacrylate ester developed to
assist adhesion to metals and plastics in peroxide cure
applications. SR-9009 features exceptionally fast cure response.
Usage levels of 5-10% are recommended.

Sartomer: Metallic Monomers for Metal Adhesion:

Metallic monomers are multifunctional acrylates and methacry-
lates that can be used to create strong adhesive bonds to metal
surfaces in a variety of applications such as:
*Rubber products
*Alkyd coatings
*Polymer concrete
*PVC plastisols
*Epoxy sealants
*Epoxy coatings

These metallic monomers are currently offered in the form
of diacrylates, dimethacrylates and monomethacrylate, namely:
*SARET-633: Metallic Diacrylate with scorch retarder
*SR-705: Metallic Diacrylate
*SR-708: Metallic Dimethacrylate
*SR-709: Metallic Monomethacrylate
*M-Cure 204: Metallic Diacrylate for Epoxy Applications

Curing Methods:
Like most acrylates and methacrylates, metallic monomers
can be cured by both free radical and Michael addition
reactions. The most common methods of initiating cure are:
*Peroxide or azo decomposition
*UV radiation
*Polyamine addition

By selecting the appropriate curing method, metallic
monomers can be used in aqueous or organic media and cured
at either ambient or elevated temperatures.

Each of the application areas in which these products
can be used to promote adhesion are discussed.

Rubber Products:
Saret 633 is a coagent for peroxide-cured rubbers.
Alkyd Coatings:
Metallic monomers also increase the adhesion of alkyd
coatings to metal substrates.
Polymer Concrete:
When SR-705 is part of the monomer system, adhesion between
the concrete and reinforcing materials is greatly increased.
PVC Plastisols:
SR-705 can be used in combination with monomers such as
trimethylolpropane trimethacrylate or with epoxies to increase
adhesion to conventional automotive substrates.
Epoxy Coatings:
M-Cure 204 also functions as a "reactive filler" in conven-
tional epoxy coatings to increase adhesion to a variety of
substrates.
Epoxy Sealants:
Acrylic monomers are used as reactive diluents in amine
cure epoxy formulations to reduce viscosity and to increase
cure speed.

Sartomer: New Tools for 2000:

New Acrylate Oligomers for Adhesives:
CN135 Low Viscosity Oligomer:
   *Adhesion
   *Low functionality
   *Low viscosity
CN137 Low Viscosity Oligomer:
   *Adhesion
   *High functionality
   *Fast cure

New Acrylate Oligomers for Inks:
CN2100 Amine-Modified Epoxy:
   *Ease of grinding
   *Quickest cure speed
   *High gloss
CN2101 Fatty Acid-Modified Epoxy:
   *Pigment grinding ease
   *High gloss
   *High Tg
CN2200 Polyester:
   *Flexible
   *Low color
   *Excellent flow
CN2201 Chlorinated Polyester:
   *Highest Tg
   *Chemical resistance
   *Fast cure speed
CN2800 Acrylated Acrylic:
   *Flexible
   *Low color
   *Excellent flow
CN2900 Aliphatic Urethane:
   *Very flexible
   *Low odor
   *Excellent flow
CN2901 Aromatic Urethane:
   *Hard film
   *Flexible
   *Low yellowing
CN2902 Aromatic Urethane:
   *Hard film
   *Flexible
   *Fast cure speed

New Acrylate Oligomers for Coatings:
CN968 Aliphatic Urethane:
   *High functionality
   *Yellowing resistance
   *Very high abrasion resistance
CN999 Aromatic Urethane:
   *Good chemical resistance
   *High surface abrasion resistance
   *Excellent wear-through resistance

Uniroyal Chemical Co., Inc.: POLYBOND Polymer Modifiers:

Polybond 1001 and 1002 Chemically Modified Polyolefins:
*Chemical coupling agents for glass and mica filled poly-
propylene systems.
*Offer good adhesion to many substrates including aluminum
and stainless steel.
*Adhesives for aluminum to aluminum and steel to steel lam-
inates.

General Properties:
Composition: Acrylic acid modified homopolymer polypropylene
Physical Form: Pellets
Melt Flow Rate (230/2.16): 40g/10 min for Polybond 1001 (ASTM
D-1238)
Melt Flow Rate (230/2.16): 20g/10 min for Polybond 1002 (ASTM
D-1238)
Density at 23C: 0.91 g/cc (ASTM D-792)
Acrylic Acid Level: 6 weight %
Melting Point: 161C (DSC)

Polybond 1009 Chemically Modified Polyolefin:
*Chemical coupling agent for glass and mica reinforced poly-
ethylene
*Provides good adhesion to a variety of substances including
aluminum, stainless steel, tin plated steel, and polyethylene

General Properties:
Composition: Acrylic acid modified high density polyethylene
Physical Form: Pellets
Melt Flow Rate (190/2.16): 5-6 g/10 min (ASTM D-1238)
Density at 23C: 0.95 g/cc (ASTM D-792)
Acrylic Acid Level: 6 weight %
Melting Point: 127C (DSC)

Polybond 1158 Chemically Modified Polyolefins:
*Outstanding adhesion to polar substrates, metal, polyprop-
ylene, and olefinic based elastomers during insert molding,
extrusion coating, and powder coating.
*Processing can be performed on conventional thermoplastic
processing equipment. Highest adhesive properties are
obtained when the substrate is heated to 160-205C during
contact with Polybond 1158.

General Properties:
Composition: Acrylic acid modified homopolymer polypropylene
Physical Form: Pellets
Melt Flow Rate (230/2.16): 10 g/10 min (ASTM D-1238)
Acrylic Acid Level: 6 weight %

Uniroyal Chemical Co., Inc.: POLYBOND Polymer Modifiers (Continued):

Polybond 3002 Chemically Modified Polyolefin:
  *Chemical coupling agent for talc, glass, and mica reinforced polypropylene giving enhanced physical and thermal properties.
  *Compatibilizer for blends such as polypropylene/polyamide and polypropylene/EVOH giving improved processability and mechanical properties.
  *Allows for processing on any conventional thermoplastic processing equipment.
  *Stabilized for both processing and long-term heat aging resistance.
General Properties:
  Composition: Maleic anhydride modified homopolymer polypropylene
  Physical Form: Pellets
  Melt Flow Rate (230/2.16): 7.0 g/10 min (ASTM D-1238)
  Density at 23C: 0.91 g/cc (ASTM D-792)
  Melting Point: 157C (DSC)

Polybond 3009 Chemically Modified Polyolefin:
  *Chemical coupling agent for glass and mica reinforced polyethylene giving improved physical and thermal properties.
  *Compatibilizer for polymer alloys of polyethylene and polar polymers, such as polyamides, to give improved processability and mechanical properties.
  *Offers good adhesion to a wide variety of substrates including aluminum, stainless steel, and polyethylene.
General Properties:
  Composition: Maleic anhydride modified high density polyethylene
  Physical Form: Pellets
  Melt Flow Rate (190/2.16): 3-6 g/10 min (ASTM D-1238)
  Density at 23C: 0.95 g/cc (ASTM D-792)
  Melting Point: 127C (DSC)

Polybond 3035 Chemically Modified Polyolefin:
  *Chemical coupling agent for glass, mica, and talc reinforced polypropylene giving enhanced physical and thermal properties
  *Compatabilizer for blends such as polypropylene/polyamide and polypropylene/EVOH to improve processability and mechanical properties.
  *Processing on any conventional thermoplastic processing equipment is feasible.
  *Stabilized for both processing and long-term heat aging resistance.
General Properties:
  Composition: Maleic anhydride modified homopolymer polypropylene
  Physical Form: Pellets
  Melt Flow Rate (190/2.16): 4.0 g/10min (ASTM D-1238)
  Density at 23C: 0.91 g/cc (ASTM D-792)
  Melting Point: 157C (DSC)

<u>**Uniroyal Chemical Co., Inc.: POLYBOND Polymer Modifiers**</u>
    (Continued):

Polybond 3109 Developmental Product Chemically Modified
    Polyolefin:
    *Chemical coupling agent for glass, mica, and ATH reinforced
    polyethylene giving improved physical and thermal properties.
    *Compatibilizer for polymer alloys of polyethylene and polar
    polymers, such as polyamides, to give improved processability
    and mechanical properties.
General Properties:
    Composition: Maleic anhydride modified linear low density
                polyethylene
    Physical Form: Pellets
    Melt Flow Rate (190/2.16): 30 g/10 min (ASTM D-1238)
    Density at 23C: 0.926 g/cc (ASTM D-792)
    Melting Point: 123C (DSC)

Polybond 3150 Chemically Modified Polyolefin:
    *Chemical coupling agent for glass, mica, and talc rein-
    forced polypropylene giving enhanced physical and thermal
    properties.
    *Compatibilizer for blends such as polypropylene/polyamide
    and polypropylene/EVOH to improve processability and mech-
    anical properties.
    *Processing on any conventional thermoplastic processing
    equipment is feasible.
    *Stabilized for both processing and long-term heat aging
    resistance.
General Properties:
    Composition: Maleic anhydride modified homopolymer polyprop-
                ylene
    Physical Form: Pellets
    Melt Flow Rate (230/2.16): 50.0g/10 min (ASTM D-1238)
    Density at 23C: 0.91 g/cc (ASTM D-792)
    Melting Point: 157C (DSC)

Polybond 3200 Chemically Modified Polyolefin:
    *Chemical coupling agent for glass, mica, and talc reinforced
    polypropylene giving enhanced physical and thermal properties
    *Compatibilizer for blends such as polypropylene/polyamide
    and polypropylene/EVOH to improve processability and mech-
    anical properties.
    *Obtain physical properties comparable to other Polybond
    products using lower addition levels.
General Properties:
    Composition: Maleic anhydride modified homopolymer polyprop-
                ylene
    Physical Form: Pellets
    Melt Flow Rate (190/2.16): 90-120 g/10 min (ASTM D-1238)
    Density at 23C: 0.91 g/cc (ASTM D-792)
    Melting Point: 157C (DSC)

Uniroyal Chemical Co., Inc.: ROYALTUF Polymer Modifiers:

Royaltuf 485 Chemically Modified Polyolefin:
  *Increases impact resistance in nylon 6, nylon 6,6, and
  other polyamide compounds.
  *Improves impact resistance in recycled polyamide materials.
General Properties:
  Composition: Maleic anhydride modified ethylene/propylene/
          non-conjugated diene elastomer
  Physical Form: Rubber Pellets
  Specific Gravity: 0.85
  Total Maleic Anhydride/Acid: 0.5%
  Mooney Viscosity: 30 (ML 1+4 at 125C)
  Iodine Number: 10
  EPDM E/P Ratio: 75/25

Royaltuf 498 Chemically Modified Polyolefin:
  *Efficient impact modifier for nylon 6, nylon 6,6, and other
  polyamide compounds
  *Particularly recommended for maximizing low temperature
  impact of nylon 6.
  *Improves impact strength of recycled polyamide products.
General Properties:
  Composition: Maleic anhydride modified ethylene/propylene/
          non-conjugated diene elastomer
  Physical Form: Rubber Pellets
  Total Maleic Anhydride/Acid: 1.0%
  Mooney Viscosity: 30 (ML 1+4 at 125C)
  Tg: -46C
  Specific Gravity: 0.89

Royaltuf 372P20 Chemically Modified Polyolefin:
  *Increases toughness of plastics such as polycarbonate, poly-
  ester/polycarbonate alloys, and SAN.
  *Outstanding weatherability and UV resistance.
General Properties:
  Composition: Styrene/acrylonitrile (SAN) modified ethylene/
          propylene/non-conjugated diene elastomer
  Physical Form: Plastic Pellets
  Specific Gravity: 0.98
  Melt Flow Rate (265/21.6): 20g/10 min.

Section II
Anti-Fogging Agents

Lonza Inc.: GLYCOLUBE AFA-1 Proprietary Anti-Fog:

Physical Properties:
    Acid Value: 2.0 Max.
    Color: 7.2 Max.
    Saponification Value: 85-95
    Hydroxy Value: 165-195
Typical Properties:
    Appearance: Amber Liquid

Suggested Applications:
    Glycolube AFA-1 imparts excellent anti-fog and anti-static
properties to PVC food wrap film and other films, far exceeding
the performance of conventional systems. Comparison of Glycolube
AFA-1 to a number of one, two, and three component anti-fog
systems has shown superior anti-fogging performance for Glycolube
AFA-1. Its highly efficient anti-fog activity also allows Glycolube AFA-1 to be used at significantly reduced loadings. The data
demonstrates that traditional two-component systems requiring
2.5 phr usage levels can be replaced by 1.5 phr of Glycolube
AFA-1.
    Glycolube AFA-1 also imparts outstanding static dissipation
properties to PVC films. A 1.5 phr loading provides static
properties sufficient to meet NFPA code 99 specifications. As
shown in the data, conventional additives are unable to provide
this level of protection even when used at higher loadings.
    Glycolube AFA-1 achieves this outstanding performance without
the adverse effects on optical properties and thermal stability
which can be caused by quaternary or amine anti-static agents.
Glycolube AFA-1 offers substantial increase in PVC thermal
stability while having no effect on optical properties.
    Glycolube AFA-1 is also an effective anti-fog agent in poly-
ethylene and polypropylene.

Recommended Usage Levels:
    Glycolube AFA-1 can be used at lower loadings than convention-
al anti-fogging/anti-static systems. Levels of 1.5-2.5 phr are
recommended for effective fog resistance in flexible PVC food
wrap formulations. Loadings as low as 1.5 phr are sufficient to
meet NFPA code 99 specifications for static dissipation.

FDA Status:
    Glycolube AFA-1 is FDA approved for use in PVC food contact
films. Its use is covered under the following paragraphs of
Title 21, US Code of Federal Regulations:
            CFR 172.854
            CFR 178.3400

Chemax, Inc.: CHEMSTAT Antifogs:

AF-476:
    Use Level % by Weight: PVC: 1.0-2.0
    Form: Liquid
    FDA: Yes
AF-640:
    Use Level % by Weight: PE/EVA: 2.0-3.0
    Form: Pellet
    FDA: Yes
AF-700/LIQ:
    Use Level % by Weight: PE/EVA: 1.0-2.0
    Form: Liquid
    FDA: No
AF-710:
    Use Level % by Weight: PE/EVA: 0.50-2.0//PVC: 3.0-4.0
    Form: Liquid
    FDA: Yes
AF-806:
    Use Level % by Weight: PE/EVA: 1.0-2.0
    Form: Pellet
    FDA: Yes
AF-1006:
    Use Level % by Weight: PP: 2.0-3.0
    Form: Pellet
    FDA: Yes

Polypropylene:
    Application: Food packaging films
    Recommended Antifog: Chemstat AF-1006
    Thermal Stability: 290C

Polyethylene/EVA:
    Application: Food packaging films (Deep freeze temperatures)
    Recommended Antifog: Chemstat AF-710
    Thermal Stability: 290C
    Application: Food packaging films (Deep freeze temperatures)
    Recommended Antifog: Chemstat AF-806
    Thermal Stability: 280C
    Application: Food & agricultural films (Warm & cold)
    Recommended Antifog: Chemstat AF-640
    Thermal Stability: 300C
    Application: 1 season agricultural films
    Recommended Antifog: Chemstat AF-700/LIQ
    Thermal Stability: 290C

PVC:
    Application: Food packaging films
    Recommended Antifog: Chemstat AF-476
    Thermal Stability: 280C
    Application: Food packaging films (Deep freeze temperatures)
    Recommended Antifog: Chemstat AF-710
    Thermal Stability: 290C

Merix Chemical Co.: MERIX Anti-Fog with Anti-Static:

A new anti-fog chemical with anti-static action has been developed by Merix Chemical Co. It is called Merix Anti-Fog.... Spray quarts and gallons are offered for industrial uses on safety goggles and shields, windows of cold chambers, instruments, freezer displays, store windows. A 4 oz. plastic container is available for in-plant distribution and as an aid for safer winter driving. Merix Anti-Fog is also used for eye-glasses, car windows, kitchen windows, and bathroom windows by industries, motels, stores, and car owners on all glass or plastic surfaces.

New Merix Anti-Fog... has de-fogging and de-staticizing chemicals acting together so that one application on glass or plastic surfaces repels fog, mist, moisture condensation as well as dust particles. It is, of course, non-flammable, also EPA registered.

Morton Plastics Additives: LUBRIOL Antifogging Agents:

AB 4192:
    Action: Internal
    Appearance: Liquid
    Density (25C): 0.970
    Food Approval: BGA
    Effective antifog system for stretch food wrap films; lubricating effect

AB 74:
    Action: Internal
    Appearance: Liquid
    Density (25C): 0.980
    Food Approval: BGA
    Effective antifog system for stretch food wrap films; lubricating effect

AB 500:
    Action: Internal
    Appearance: Liquid
    Density (25C): 0.990
    Food Approval: BGA, FDA
    Antifog system for stretch food wrap films; FDA approved

Patco Additives Division: PATIONIC Antifogging Agents:

Pationic 907:
Pationic 907 is an unsaturated, distilled (95% minimum, monoester) monoglyceride derived from vegetable oils.
Typical Applications:
Rigid PVC compound internal lubricant
Antifogging Agent:
Flexible PVC compounds-95% monoester content provides better compatibility between the hydrophilic component of an antifog formulation and the PVC film or sheet, while wash-off resistance is maintained. PVC Compound thermal stability is enhanced and color development is minimized.
Polyolefins-Primary additive that imparts moisture wettability on the surface. Clarity is maintained and color development is minimized.

Pationic 951:
Pationic 951 is an unsaturated distilled monoglyceride derived from vegetable oils having a high oleic content.
Typical Applications:
Rigid PVC compound internal lubricant
Antifogging Agent:
Flexible PVC compounds-95% monoester content provides compatibility between the hydrophilic component of an antifog formulation and the film or sheet, while wash-off resistance is maintained. PVC compound thermal stability is enhanced and color development is minimized.
Polyolefins-Primary additive that imparts moisture wettability on the surface. Clarity is maintained and color development is minimized. Pationic 951 has excellent oxidative stability due to its high level of monounsaturation.

Pationic 1001:
Pationic 1001 is a glycerol mono oleate product. It is derived from natural fats and oils, and incorporates a surfactant for aqueous self-emulsification.
Typical Current Applications:
Pationic 1001 is an antistat and antifog for polyvinyl chloride, polyolefins, and styrenics. It can be formulated internally for surface migration, or is suitable for external application. It is a semi-solid at room temperature and incorporates a surfactant for the easy preparation of stable, aqueous emulsions. Pationic 1001 is also a lubricant and additive dispersing agent.

Patco Additives Division: PATIONIC Antifogging Agents (Continued):

Pationic 1004:
Pationic 1004 is a liquid glycerol mono oleate product. It is derived from natural fats and oils, and incorporates a surfactant for aqueous self-emulsification.
Typical Current Applications:
Pationic 1004 is an antistat and antifog for polyvinyl chloride, polyolefins, and styrenics. It can be formulated internally for surface migration, or is suitable for external application. It is a liquid at room temperature and incorporates a surfactant for the easy preparation of stable, aqueous emulsions. Pationic 1004 is also a lubricant and additive dispersing agent.

Pationic 1033:
Pationic 1033 is a glycerol mono oleate product with a 52%, minimum, alpha-monoester content. It is derived from natural fats and oils.
Typical Current Applications:
Pationic 1033 is used as an additive to PVC, polyolefins, and styrenics. The product is an excellent internal lubricant for PVC that helps maintain clarity and early color hold. It is also a common component in PVC antifogging formulations to promote better compatibility between the resin and the hydrophilic component of the formulation. In polyolefins, Pationic 1033 functions as a flow modifier and an additive dispersing agent.

Pationic 1064:
Pationic 1064 is a glycerol mono oleate product with a 42%, minimum, alpha-monoester content. It is derived from natural vegetable oils. A Kosher grade is available upon request.
Typical Current Applications:
Pationic 1064 is used as an additive to PVC, polyolefins, and styrenics. The product is an excellent internal lubricant for PVC that helps maintain clarity and early color hold. It is also a common component in PVC antifogging formulations to promote better compatibility between the resin and the hydrophilic component of the formulation. In polyolefins, Pationic 1064 functions as a flow modifier and an additive dispersing agent.

**Patco Additives Division: PATIONIC Antifogging Agents
(Continued):**

Pationic 1068:
Pationic 1068 is a glycerol mono oleate product with a 52%,
minimum, alpha-monoester content. It is derived from natural
fats and oils. Pationic 1068 is very high in mono-unsaturates.
Typical Current Applications:
Pationic 1068 is used as an additive to PVC and polyolefins.
The product is an excellent internal lubricant for PVC that
helps maintain clarity and early color hold. It is also an
antifog component in flexible PVC films. In polyolefins,
Pationic 1068 functions as a flow modifier and an additive
dispersing agent.

Pationic 1074:
Pationic 1074 is a glycerol mono oleate product with a 42%,
minimum, alpha-monoester content. It is derived from natural
fats and oils. A Kosher grade is available upon request.
Typical Current Applications:
Pationic 1074 is used as an additive to PVC, polyolefins,
and styrenics. The product is an excellent internal lubricant
for PVC that helps maintain clarity and early color hold. It is
also a common component in PVC antifogging formulations to
promote better compatibility between the resin and the hydrophilic component of the formulation. In polyolefins, Pationic
1074 functions as a flow modifier and an additive dispersing
agent.

Pationic 1083:
Pationic 1083 is a glycerol mono oleate product with a 52%,
minimum, alpha-monoester content. It is derived from natural
fats and oils.
Typical Current Applications:
Pationic 1083 is used as an additive to PVC, polyolefins,
and styrenics. The product is an excellent internal lubricant
for PVC that helps maintain clarity and early color hold. It
is also a common component in PVC antifogging formulations
to promote better compatibility between the resin and the
hydrophilic component of the formulation. In polyolefins,
Pationic 1083 functions as a flow modifier and an additive
dispersing agent.

Patco Additives Division: PATIONIC Antifogging Agents (Continued):

Pationic 1087:
Pationic 1087 is a glycerol mono oleate product with a 52%, minimum alpha-monoester content. It is derived from natural fats and oils.
Typical Current Applications:
Pationic 1087 is used as an additive to PVC and polyolefins. The product is an excellent internal lubricant for PVC that helps maintain clarity and early color hold. It is also a common component in PVC antifogging formulations. In polyolefins, Pationic 1087 functions as a flow modifier and an additive dispersing agent. High monounsaturation makes for good oxidative stability.

Pationic 1530:
Pationic 1530 is a complex glycerol ester derived from natural fats and oils. Ambient Pationic 1530 is a clear liquid, resistant to settling. It stays fluid and pumpable at lower temperatures, where many glycerol esters solidify or turn to paste.
Typical Applications:
The primary use of Pationic 1530 is an antifog additive for flexible PVC or polyolefin film. It can be used alone or in combination with other surfactants. In addition, Pationic 1530 can be used as a PVC lubricant, a polyolefin mold release, or an antistat in a variety of polymers.

Patco Additives Division: PATIONIC Antifogging Agents
(Continued):

Pationic 1087:
    Pationic 1087 is a glycerol mono oleate product with a 52%,
minimum alpha-monoester content. It is derived from natural
fats and oils.
Typical Current Applications:
    Pationic 1087 is used as an additive to PVC and polyolefins.
The product is an excellent internal lubricant for PVC that
helps maintain clarity and early color hold. It is also a
common component in PVC antifogging formulations. In polyolefins,
Pationic 1087 functions as a flow modifier and an additive
dispersing agent. High monounsaturation makes for good oxidative stability.

Pationic 1530:
    Pationic 1530 is a complex glycerol ester derived from
natural fats and oils. Ambient Pationic 1530 is a clear liquid,
resistant to settling. It stays fluid and pumpable at lower
temperatures, where many glycerol esters solidify or turn to
paste.
Typical Applications:
    The primary use of Pationic 1530 is used an antifog add-
itive for flexible PVC or polyolefin film. It can be used alone
or in combination with other surfactants. In addition, Pationic
1530 can be used as a PVC lubricant, a polyolefin mold release,
or an antistat in a variety of polymers.

Witco Corp.: Additives for Polyolefins: Recommended Uses:
Antifog & Antistat:

Atmos 150*:
FDA Sanctioned: Yes
LDPE/LLDPE/Polypropylene

Atmos 300*:
FDA Sanctioned: Yes
EVA Modified PE/LDPE/LLDPE/Polypropylene

Atmul 84*:
FDA Sanctioned: Yes
LDPE/LLDPE/Polypropylene

Atmul 695*:
FDA Sanctioned: Yes
EVA Modified PE/LDPE/LLDPE/Polypropylene

Dimul S*:
FDA Sanctioned: Yes
EVA Modified PE/LDPE/LLDPE/Polypropylene

Kemamine AS-650:
FDA Sanctioned: Yes
EVA Modified PE/HDPE/LDPE/LLDPE/Polypropylene/UHMWPE

Kemamine AS-974:
FDA Sanctioned: Yes
EVA Modified PE/HDPE/LDPE/LLDPE/Polypropylene

Kemamine AS-974/1:
FDA Sanctioned: Yes
EVA Modified PE/HDPE/LDPE/LLDPE/Polypropylene

Kemamine AS-989:
FDA Sanctioned: Yes
EVA Modified PE/HDPE/LDPE/LLDPE/Polypropylene

Kemamine AS-990:
FDA Sanctioned: Yes
EVA Modified PE/HDPE//LDPE/LLDPE/Polypropylene

MoldPro 873:
FDA Sanctioned: Yes
EVA Modified PE/HDPE/LDPE/LLDPE/Polypropylene

*Kosher grade available

Section III
Antioxidants

Bayer Corp.: VULCANOX/VULCANOL Antioxidants/Antiozonants:

Vulcanox 3100:
 Staining Antioxidant
 Chemical Composition: Blend of diaryl p-phenylene diamines
 Properties:
 Form: Brown flakes
 Active substance content (% min.): 80
 Specific gravity at 73F/23C (g/cm3): 1.20
 Melting point (min.): 203F/95C
 Discoloration: Discoloring and staining
 Application:
 Vulcanox 3100 is a powerful, persistent antioxidant and anti-
flexcracking agent for most diene rubbers (NR, IR, SBR, NBR),
imparting moderate antiozonant properties. Ozone resistance
can be improved by the addition of microcrystalline waxes.
In polychloroprene (CR) excellent protection against ozone
attack is provided. Vulcanox 3100 is advantageously used in
conjunction with Vulcanox 4010 NA (IPPD) or Vulkanox 4020 (6PPD)
antidegradants.
 Typical application areas include tires (protection against
groove cracking), V-belts, axle boots, buffers, engine mounts,
conveyor belts and other mechanical goods.

Vulkanox 4010 NA/LG:
 Staining Antiozonant
 Chemical Composition: N-isopropyl-N'-phenyl-p-phenylene
                       diamine (IPPD)
 Molecular weight: 226
 Properties:
 Form: Brown granules
 Active substance content (% minimum): 95
 Specific gravity at 73F/23C (g/cm3): 1.07
 Melting point (min.): 169F/76C
 Discoloration: Discoloring and staining

Application:
 Vulkanox 4010 NA/LG is a staining and discoloring antiozonant
and antioxidant protecting rubber goods against ozone attack,
oxidation, heat aging, flexcracking and rubber poisons. It is
suitable for natural and synthetic rubbers.
 Vulkanox 4010 NA/LG antiozonant excels in antiflexcracking
properties and is used in tires and mechanical goods subjected
to dynamic stress, e.g. conveyor belts, hoses, spring components
and elastic couplings. In static applications and in cables and
seals, its main function is resistance to ozone cracking, which
can be further improved by the simultaneous use of an ozone
protective wax. Use should be restricted to dark colored rubber
articles where staining and discoloration are of no concern.

Bayer Corp.: VULKANOX/VULCANOL Antioxidants/Antiozonants (Continued):

Vulkanox 4020:
   Staining Antiozonant
   Chemical Composition: N-(1,3-dimethyl butyl) N'-phenyl-p-
                         phenylene diamine (6PPD)
   Molecular Weight: 268
   Properties:
      Form: Vulkanox 4020 LG: Brown to brown-purple pastilles
            Vulkanox 4020 Fused: Liquid
      Active substance content (% min.): 97
      Specific gravity: 1.02
      Discoloration: Discoloring and staining
      Solidification Point: 113F/45C min.
   Application:
      Vulkanox 4020 antiozonant is a staining and discoloring antio-
   zonant and antioxidant which helps protect rubber goods against
   ozone attack, oxidation, heat aging, flexcracking, and rubber
   poisons. It is suitable for natural and synthetic rubbers.

Vulkanox HS:
   Staining Antiozonant
   Chemical Composition: Polymerized 1,2-dihydro-2,2,4-trimethyl
                         quinoline (TMQ)
   Molecular Weight: 172
   Properties:
      Form: Vulkanox HS, LG: Yellow to tan colored granules
            Vulkanox HS, Powder: Yellow to tan colored powder
      Alkalinity index (%): 520-580
      Density (g/cm3): 1.1
      Softening point (Mettler): 83C min (Powder)
                                 85-95C (Granules)
      Discoloration: Discoloring and staining
   Application:
      Vulkanox HS is an antioxidant with relatively weak staining
   and discoloration characteristics. It provides outstanding heat
   protection in natural rubber (NR), synthetic polyisoprene (IR),
   polybutadiene (BR), styrene butadiene rubber (SBR), nitrile
   rubber (NBR) and ethylene propylene rubber (EPDM), but is less
   suitable for polychloroprene (CR). Its performance is improved
   further in combination with Vulkanox MB-2 antioxidant, which is
   especially effective in thiuram cured articles with low levels
   or sulfur or without sulfur. Vulkanox HS antioxidant also yields
   good protection against rubber poisons.

Bayer Corp.: VULCANOX/VULKANOL Antioxidants/Antiozonants (Continued):

Vulkanox KB:
"Non-Staining" Antioxidant
Chemical Composition: 2,6-di-tertiary-butyl-p-cresol (BHT)
Molecular Weight: 220
Properties:
Form: Colorless crystals
Active substance (% min.): 99.8
Solidification point (min.): 156.5F/69.2C
Specific gravity at 73F/23C (g/cm3): 1.03
Application:
Vulkanox KB is a "non-discoloring antioxidant" and "non-staining" antioxidant for light colored and transparent rubber goods based on natural and synthetic rubbers. Antioxidant properties can be improved in combination with Vulkanox MB-2 antioxidant.
Typical applications include proofed goods, toys, footwear, threads, floor covering, mats and latex articles.

Vulkanox MB-2/MGC:
"Non-Staining" Antioxidant
Chemical Composition: Blend of 4- and 5- methylmercapto
                      benzimidazole, mineral oil coated
Molecular Weight: 164
Properties:
Form: Yellow-white micro-granules
Active substance content (% min): 96.5
Specific gravity at 73F/23C (g/cm3): 1.25
Oil content (%): 1.5-2.0
Application:
Vulkanox MB-2/MGC is a "non-discoloring" and "non-staining" antioxidant for natural and synthetic rubbers. It provides excellent resistance against heat and oxygen aging, especially when combined with amine or phenolic antioxidants. The outstanding performance in conjunction with other antioxidants (e.g. Vulkanox 4010 NA; 4020, HS, BKF antioxidants) is due to a synergistic effect, yielding superior results to any of the separate constituents. Maximum resistance is provided in thiuram and dithiocarbamate curing systems containing little or no sulfur. Additions of Vulkanox MB-2/MGC to polychloroprene (CR) impart a significant improvement in flexcracking.

Bayer Corp.: VULKANOX/VULKANOL Antioxidants/Antiozonants (Continued):

Vulkanox ZMB 2:
   Non staining antioxidant
   Chemical Composition: Zinc salt of a blend of 4- and 5-methyl-
                          2-mercaptobenzimidazole
   Molecular Weight: 392
   International Abbreviation: ZMMBI
Product Description:
   Supply Form: Vulkanox ZMB 2/C 5: White to beige powder,
                                     mineral oil coated
   Zinc content (%): 15.5
   Density at 20 (g/cm3): 1.5
   Oil content (%): 5.0
   Discoloration: non-staining, non-discoloring
Application:
   Vulkanox ZMB 2 is a non-discoloring and non-staining anti-
oxidant for natural and synthetic rubbers. It provides excellent
protection against heat and oxygen aging, especially when com-
bined with amine or phenol type antioxidants. Its performance
in conjunction with other antioxidants is synergistic. In
polychloroprene, a significant improvement in flex-cracking is
observed.

Vulkanol 85:
   Ether thioether
   Description: Supply form: Nearly colorless to yellow liquid
Properties:
   Density at 20C: 1.045+-0.025 g/cm3
   Refractive Index at 20C: 1.4705+-0.0025
   Viscosity at 20C: 60+-2 mPas
Other Product Features:
   Volatile matter: 2.5%

Ciba Specialty Chemicals: Additives for Polyolefins:
Antioxidants (AO) & Processing Stabilizers:

Primary Antioxidants:
Irganox 1010:
   High molecular weight, phenolic AO with low volatility;
extends long-term thermal stability (LTTS)
Irganox 1035:
   Phenolic AO for cross-linked or carbon black containing sys-
tems (i.e. cable compounds)
Irganox 1076:
   Highly compatible, low-color phenolic AO for PE; available
in preferred product forms.
Irganox 1081:
   High performance, highly compatible AO for cross-linked PE
(i.e. high voltage powder cable)
Irganox 1330:
   Extraction resistant phenolic AO with low water carryover for
film and tape
Irganox 1425 WL:
   Phenolic AO with high extraction resistance; polyethylene
wax carrier
Irganox 3114:
   Highly effective, non-discoloring phenolic antioxidant, high
extraction resistance
Irganox E 201:
   Vitamin E, a high efficiency phenolic AO with consumer appeal
and excellent MF stability
Irganox E 217:
   Granular, dispersible solid with GRAS coadditives for food
and medical packaging
Irganox E 218:
   Low dust Vitamin E blend with calcium stearate for food and
medical packaging
Irganox MD 1024:
   High performance phenolic AO for applications requiring metal
deactivation (Cu passivator)

Secondary Antioxidants:
Irgafos 168:
   Hydrolytically stable phosphite, synergy with primary AO's,
maintains color properties.
Irgafos 38:
   High performance, highly compatible phosphite with low melt
viscosity
Irgafos P-EPQ:
   High performance, highly compatible phosphonite with low melt
viscosity; offers low color
Irganox PS 800:
   Extends long-term thermal stability as a synergist for pheno-
lic antioxidants
Irganox PS 802:
   Extends long-term thermal stability as a synergist for pheno-
lic antioxidants

Ciba Specialty Chemicals: Additives for Polyolefins:
Antioxidants (AO) & Processing Stabilizer Blends:

Irganox B 1411:
    Synergistic phenol/phosphite blend mainly for PP applications
Irganox B 1412:
    Synergistic phenol/phosphite blend with higher phosphite
levels mainly for PP applications
Irganox B 215:
    Synergistic, convenient phenol/phosphite blend for a balance
of stabilization requirements
Irganox B 220:
    Synergistic blend; high phosphite content for demanding
processing conditions
Irganox B 225:
    Synergistic, convenient blend with high phenolic AO content;
extends LTTS
Irganox B 311:
    Synergistic phenol/phosphite blend for low water carry-over
Irganox B 313:
    Synergistic blend with high phenolic content for increased
LTTS, low water carry-over
Irganox B 501 W:
    Synergistic blend of phenol/phosphite for PP fiber
Irganox B 561:
    Synergistic phenol/phosphite blend for demanding processing
conditions
Irganox B 900:
    Synergistic phenol/phosphite blend with high phosphite levels
mainly for PE applications
Irganox B 921:
    Synergistic phenol/phosphite blend with high phosphite levels
mainly for PE applications
Irganox HP 2215:
    High performance, cost-effective stabilizer system with added
processing stability
Irganox HP 2225:
    High performance, cost-effective stabilizer system for PP and
demanding PE applications
Irganox HP 2251:
    High performance, cost-effective stabilizer, additional pheno-
lic AO for LTTS
Irganox HP 2411:
    High performance, cost-effective stabilizer system primarily
for molding applications and fibers
Irganox HP 2921:
    High performance, cost-effective stabilizer system for demand-
ing PE applications
Irganox XP 420:
    Highly efficient system for PE applications requiring melt
flow control and low color
Irganox XP 490: Highly efficient processing stabilization
Irganox XP 620: Highly efficient stabilization for PP and HDPE
Irganox XP 621: Highly efficient stabilization for PP and HDPE

Cytec Industries Inc.: CYANOX Antioxidants:

For processing stabilization, Cytec's Cyanox phenolic antioxidants and phenolic phosphite blends offer excellent protection
against thermo-oxidative degradation and discoloration of polymers. The Cyanox thioester antioxidants are efficient synergists
used in phenolic antioxidant polymer applications requiring good
thermal aging properties.

Phenolic Antioxidants:
425:
    Chemical Formula: C25H36O2
    CAS Number: 000088-24-4
    A bisphenol recommended for use in rubber-modified plastics.

2246:
    Chemical Formula: C23H32O2
    CAS Number: 000119-47-1
    A bisphenol recommended for use in ABS and SBR latexes.

1741:
    Chemical Formula: C48H69N3O6
    CAS Number: 27676-62-6
    A high molecular weight hindered phenol recommended for
high performance polypropylene fiber applications.

1790:
    Chemical Formula: C42H57N3O6
    CAS Number: 040601-76-1
    A high performance, gas fade-resistant trisphenol, character-
ized by low volatility, recommended for polyolefins and polyurethane elastomers.

2777:
    Chemical Formula: Mixture/Phosphite
    CAS Number: 040601-76-1/031570-04-0
    A blend of Cyanox 1790 antioxidant and phosphite providing
outstanding melt processing stabilization of polyolefins without
pinking problems.

Cytec Industries Inc.: CYANOX Antioxidants

Thioester Antioxidants:
LTDP:
Chemical Formula: C30H58O4S
CAS Number: 000123-28-4
A sulfur synergist for phenolic antioxidants recommended primarily for in-service thermal stabilization of polyethylenes.

STDP:
Chemical Formula: C42H82O4S
CAS Number: 000693-36-7
A sulfur synergist for phenolic antioxidants recommended primarily for in-service thermal stabilization of polypropylenes.

1212:
Chemical Formula: C36H70O4S
CAS Number: 000123-28-4/000693-36-7/013103-52-1
A sulfur synergist for phenolic antioxidants recommended primarily for in-service thermal stabilization of polypropylenes.

711:
Chemical Formula: C32H62O4S
CAS Number: 010595-72-9
A liquid sulfur synergist for phenolic antioxidants, recommended for ABS and SBR latex applications.

Fairmount Chemical Co., Inc.: MIXXIM AO-30 Antioxidant:

1,1,3-tris-(2-methyl-4-hydroxy-5-tert-butylphenyl)-butane

MIXXIM AO-30 is broadly approved by the FDA for use as an antioxidant in products that are in contact with food. It is a non-staining, high molecular weight, hindered phenolic compound which prevents thermal and oxidative degradation.

Applications:
Mixxim AO-30 provides excellent protection against oxidation when used at levels of 0.02 to 1%, depending on the polymer, application and process used. It is recommended as an anti-oxidant in various plastics applications that involve demanding processing and end use conditions: Polyolefins * Styrenics * Rubber Modified Styrenics (such as ABS, SAN) * Block Copolymers * Elastomers * PVC * Engineering Plastics * Hot Melt Additives

Advantages:
*Excellent heat, processing and color stability
*High synergistic with phosphites and thioesters
*Strong resistance to extraction by foodstuffs, hot water or detergents
*Broadest FDA clearance given for any high molecular weight phenolic
*Low level of use, cost effective
*Soluble in a wide range of solvents
*Non staining
*Flows freely for easy dispersion into polymers

Synergistic Performance:
The overall performance and cost efficiency of Mixxim AO-30 may be enhanced when it is used in conjunction with phosphites and/or thioesters.
Resistance to Extraction:
Mixxim AO-30 has high resistance to extraction giving the end product long term protection against degradation.
Resistance to Extraction in Polypropylene

Typical Properties:
Appearance: White flowable powder
Melting Point: 185-190C
Moisture: Karl Fisher 1%, max.
IR Spectrum: To match standard

Structural Formula:
Molecular Formula: C37H52O3
Molecular Weight: 544
CAS Registry No.: 1843-03-4

GE Specialty Chemicals: WESTON & ULTRANOX Antioxidants:

Ultranox 626, 626A, 627A Phosphite Antioxidants:
Bis (2,4-di-t-butylphenyl) Pentaerythritol Diphosphite
High-performance solid organophosphite antioxidants which
are extremely cost-efficient. They are very effective in the
stabilization of polyolefins, polyesters, styrenics, engineering
thermoplastics, PVC, elastomers, and adhesives.
Ultranox 626, 626A, 627A phosphite antioxidants contribute
significantly in achieving:
* Excellent color stability during compounding, fabrication,
  and end use.
* A reduction in polymer degradation during processing.
* Improved gas fading tendencies in many applications.
* Higher phosphorous content than any other solid phosphite.
* Synergism when used with light stabilizers such as
  benzophenones and benzotriazoles.
* High performance at low loading levels for more cost
  effective formulations.

Ultranox 668 Phosphite Antioxidant:
Applications for Ultranox 668 include polyolefins, polyesters,
engineering thermoplastics, and PVC.
Ultranox 668 is sanctioned in many countries for certain food
packaging applications.

Ultranox 641 Phosphite Antioxidant:
High performance solid phosphite antioxidant. It offers high
stabilization, excellent hydrocarbon solubility, and excellent
hydrolytic stability.
Applications for Ultranox 641 include polyolefins - such as
films and fibers, styrenics, engineering thermoplastics, PVC
elastomers and adhesives.

Ultranox 210 Phenolic Antioxidant:
A high molecular weight, multifunctional phenolic antioxidant
for improving processing stability and long term heat aging in
a variety of polymers. Specific applications include styrenics,
polyolefins, PVC, urethanes, acrylics, adhesives and elastomers.

Ultranox 276 Phenolic Antioxidant:
A high molecular weight phenolic antioxidant for improving
processing stability in a variety of polymers. Specific applications include styrenics, polyolefins, PVC, urethane and acrylic
coatings, adhesives and elastomers. It is also an effective
replacement for BHT in polyolefins.

GE Specialty Chemicals: WESTON & ULTRANOX Antioxidants (Continued):

Ultranox Antioxidant Blends:
Blends of Ultranox phosphite and phenolic antioxidants in a single package. Ultranox antioxidant blends reduce handling, and provide consistent component quality and mix ratios. They are available as free flowing solids with controlled particle size.
GE Specialty Chemicals can also provide custom blends.

Weston 618F, 619F Phosphites:
Distearyl Pentaerythritol Diphosphites
Highly effective color and molecular weight stabilizers for polyolefins, polyesters, elastomers, styrenics, engineering thermoplastics, and adhesive formulations. They also contribute to the thermal and UV stability of these polymers. Some benefits include:
* Excellent color stability during processing, fabrication, and end use.
* Polymer protection from degradation during processing.
* Reduced discoloration of:
  - polymers containing pigments such as TiO_2.
  - polymers that are stabilized with a sulfur-containing nickelorganic complex and are processed at high temperatures.
* Synergism when used with light stabilizers such as benzophenones and benzotriazoles.
* High performance at low loading levels for more cost effective formulations.
Weston 618F and 619F phosphites are sanctioned for food contact applications in many countries.

Weston DHDP:
Poly(dipropyleneglycol) Phenyl Phosphite
A high molecular weight phosphite which has a relatively high phosphorus content. Useful as a secondary stabilizer for PVC resins where regulatory approval for food is not required.

Weston DOPI:
Diisooctyl Phosphite
An alkyl phosphite which functions as a secondary stabilizer for PVC. As a lubricant additive, Weston DOPI phosphite improves antifriction and antiwear characteristics. It can also be used as a color and molecular weight stabilizer for polypropylene when regulatory approval for food contact is not required.

GE Specialty Chemicals: WESTON & ULTRANOX Antioxidants (Continued):

Weston DPDP:
Diphenyl Isodecyl Phosphite
An alkyl-aryl phosphite which is used as a color and processing stabilizer in polycarbonate, polyurethanes, ABS polymers,
and coatings where regulatory approval for food contact is not
required. It is also used as a secondary stabilizer in polyvinyl chloride.

Weston DPP:
Diphenyl Phosphite
A dialkyl phosphite which can be used to improve the color
of unsaturated polyesters. In some polymers, Weston DPP phosphite can be used as a stabilizer in applications where regulatory approval for food contact is not required. It is also
used as a sceondary stabilizer to improve the light and color
stability and to prevent plate-out in polyvinyl chloride. This
product has a high phosphorus content and a relatively low
molecular weight.

Weston EGTPP:
Epoxy Grade Triphenyl Phosphite
Used in epoxy resin systems as a reactive diluent. It can
be used (1) as a diluent without extra filler to achieve a
more economical compound and (2) to reduce the viscosity of
the epoxy resin, making it more manageable as additional
amounts of filler are added.

Weston ODPP:
Diphenyl Isooctyl Phosphite
An alkyl-aryl phosphite which can be used as a color and
processing stabilizer in ABS, polycarbonate, polyurethane,
coatings, and PET fiber where regulatory approval for food
contact is not required. It can also be used as a secondary
stabilizer to improve color and heat stability in PVC.

Weston PDDP:
Phenyl Diisodecyl Phosphite
An alkyl-aryl phosphite which can be used as a color and
processing stabilizer in ABS, polycarbonate, polyurethane,
coatings, and PET fiber where regulatory approval for food
contact is not required. It can also be used as a secondary
stabilizer to improve color and heat stability in PVC.

GE Specialty Chemicals: WESTON & ULTRANOX Antioxidants (Continued):

Weston PNPG:
    Phenyl Neopentylene Glycol Phosphite
    A specialty alkyl-aryl phosphite with a very high phosphorus content. It is an effective secondary stabilizer for a variety of polymers where regulatory approval for food contact is not required.

Weston PTP:
    Heptakis (dipropyleneglycol) Triphosphite
    A specialty product which is useful when a high molecular weight phosphite is needed. It has a free hydroxyl number of approximately 275 mg KOH/g. Weston PTP phosphite in foamed polyurethanes, controls color development and prevents burn scorching.

Weston TDP:
    Triisodecyl Phosphite
    A trialkyl phosphite which is used as a secondary stabilizer in ABS, polyethylene terephthalate, PET fibers, coatings, lubricants, PVC, and polyurethanes. It can also be used in polyolefins where regulatory approval for food contact is not required.

Weston THOP:
    Tetraphenyl Dipropyleneglycol Diphosphite
    A specialty diphosphite which is useful as a secondary stabilizer in many polymers, including PVC where regulatory approval for food contact is not required.

Weston TIOP:
    Triisooctyl Phosphite
    A trialkyl phosphite. It can be used in acrylics to improve color and molecular weight retention during processing. It also provides stability to nylon, unsaturated polyester, and PVC. When Weston TIOP phosphite is used in lubricants, it functions as a sulfur deactivator. It can also be used to improve antifriction and antiwear characteristics.

Weston TLP:
    Trilauryl Phosphite
    Can be used as a lubricant additive. Also functions as a sulfur deactivator. Weston TLP phosphite may be used as a stabilizer in PVC, polyester fibers and in polypropylene where regulatory approval for food contact is not required.

GE Specialty Chemicals: WESTON & ULTRANOX Antioxidants (Continued):

Weston TLTTP:
A unique phosphite which contains sulfur. It can be used as an effective heat and thermal stabilizer for a variety of polymers. It is particularly effective in polyolefins, where it acts to improve processing and UV stability. Can also be used as an antioxidant and a sulfur deactivator in lubricants while improving antifriction and antiwear characteristics.

Weston TNPP, 399:
Tris (Nonylphenyl) Phosphites
* High purity   * Low color
* Low nonylphenol content
Versatile phosphite stabilizers which are useful in a large number of polymers such as HDPE, LLDPE, SBR, ABS, PVC and others. They are sanctioned in many countries for food packaging applications.

Weston TPP:
Triphenyl Phosphite
A versatile aryl phosphite which can be used as a stabilizer in many types of polymers including: adhesives, styrenics, engineering thermoplastics, polyesters (to regulate viscosity and improve color stability), polyolefins (as a catalyst adjuvant), polyurethanes (to prevent scorching and improve color stability), coatings, epoxies, and PVC where regulatory approval in food contact is not required.

Weston 430:
A specialty alkyl phosphite. It has a hydroxyl number of 395 mg KOH/g. It is an effective heat, color, and viscosity stabilizer for a variety of polymers including polyester fibers and polyurethanes.

Weston 439:
Poly 4,4' Isopropylidenediphenol--C12-15 Alcohol Phosphite
Sanctioned by the FDA for use in rigid PVC and rigid vinyl chloride copolymers in food packaging applications.
Weston 439 phosphite is used to improve the color and light stability and clarity of PVC and vinyl chloride copolymers.

Weston 494:
Diisooctyl Octylphenyl Phosphite
A specialty alkyl-aryl phosphite useful in providing heat and color stability to a variety of polymers. It can be used in PVC as a color and light stabilizer.

Weston 600:
Diisodecyl Pentaerythritol Diphosphite
Useful as a stabilizer for a variety of polymer systems, including polycarbonate.

Goodyear Chemical: WINGSTAY Antioxidants:

Non-Staining:
Wingstay L:
Butylated reaction product of p-cresol and DCPD
Physical Form: Off-white powder or tan flakes
An outstanding non-staining antioxidant. Highly active, low
water solubility and virtually non-volatile giving long term
protection under severe conditions. Extremely effective in latex
compounds and due to its high activity can be used at low concentrations.

Wingstay S:
Styrenated phenol
Physical Form: Straw colored liquid or 70% active powder
A low volatility, non-discoloring, general purpose anti-
oxidant which offers good resistance to hardening and general
surface deterioration of rubber and latex articles.

Wingstay T:
Butylated acetylated phenol
Physical Form: Straw colored liquid
A general purpose non-staining antioxidant particularly useful in stabilizing SBR rubber. Affords good activity with moderate volatility characteristics. Imparts excellent resistance to
ultraviolet light discoloration.

Wingstay C:
Arylated alkylated phenol
Physical Form: Amber colored liquid
A low volatility, non-staining and non-discoloring antioxidant
designed for stabilization of SBR. Provides long term stability
while resisting hydrolysis in the presence of water.

Wingstay 29:
Styrenated diphenylamine
Physical Form: Straw colored--amber liquid or 70% active
powder
An extremely active, low volatility antioxidant. Protects
aganst degradation due to heat, light and oxygen. It is non-
blooming, non-hydrolysable and has no effect on cure rates.

Wingstay SN-1:
1:11-(3,6,9-trioxaundecyl) propionate
Physical Form: Colored semi solid or 70% active powder
A low volatility, non-discoloring secondary antioxidant.
It is highly active when used in combination with primary
amine or phenolic antioxidants.

Goodyear Chemicals: WINGSTAY Antioxidants (Continued):

Non-Staining (Continued):
Wingstay K:
　　Reaction product of p-nonyl phenol dodecanethiol and form-
aldehyde
　　Physical Form: A light amber liquid
　　A highly active non-volatile antioxidant which is non-staining
and non-discoloring. It is recommended for use as a raw polymer
stabilizer.

Staining:
Wingstay 100:
　　Mixed diaryl p-phenylene diamines
　　Physical Form: Blue brown flakes
　　A powerful and persistent antioxidant/antiozonant having
excellent anti-flex and anti-groove cracking characteristics
coupled with long term antiozonant properties. Not leached
out by water.

Wingstay 100 AZ:
　　Mixed diaryl p-phenylene diamines
　　Physical Form: Blue brown flakes
　　A powerful and persistent antioxidant/antiozonant designed
especially for use in polychloroprene. Specifically effective
in compounds which are prone to bin cure by reducing the potential of generating scrap due to premature crosslinking.

Wingstay 200:
　　Mixed diaryl p-phenylene diamines
　　Physical Form: Dark brown semi-solid
　　Similar to Wingstay 100 and used mainly as an emulsion
polymer stabilizer or where a liquid antiozonant is preferred.
Very stable to acid coagulation of emulsion polymers.

Antioxidant Monomers:
POLYSTAY, POLYSTAY AA-1R:
　　N-(A-Anilinephenyl) metharylamide
　　Physical Properties: A crystalline powder with a purity of
>99%
　　Polymerizable in emulsion, solution, and bulk/suspension
systems. Functions in all free radical polymerizations. AZO
initiators are preferred. May be incorporated at conventional
levels in the total polymer, added as a masterbatch when polymerized at a high concentration, or grafted into polymers.
Masterbatches may be used for dry rubber or latex compounding.
Synergizes with secondary antioxidants. When polymer bound, it
is non-volatile, non-extractable, non-migratory, and non-
staining.

Great Lakes Chemical Corp.: Antioxidants:

Phenols:
Lowinox BHT:
  2,6-Di-tert.butyl-4-methyl-phenol
  MW: 220   Melting Range C: 68-70
  CAS No. 128-37-0   Physical Form: Powder

Lowinox 22M46:
  2,2'-Methylene-bis-(4-methyl-6-tert-butyl-phenol)
  MW: 341   Melting Range C: 127-129
  CAS No. 119-47-1   Physical Form: Powder

Lowinox 44B25:
  4,4'-Butylidene-bis-(2-tert-butyl-5-methyl-phenol)
  MW: 383   Melting Range C: 208-210
  CAS No. 85-60-9   Physical Form: Powder

Lowinox TBM-6:
  4,4'-Thio-bis-(2-tert-butyl-5-methyl-phenol)
  MW: 358   Melting Range C: 160-164
  CAS No. 90-69-5   Physical Form: Powder, NDB

Lowinox TBP-6:
  2,2'-Thio-bis-(6-tert-butyl-4-methyl-phenol)
  MW: 358   Melting Range C: 83-85
  CAS No. 90-66-4   Physical Form: Powder

Lowinox AH25:
  2,5-Di-tert. amyl-hydroquinone
  MW: 250   Melting Range C: 177-179
  CAS No. 79-74-3   Physical Form: Powder

Lowinox CPL:
  Polymeric sterically hindered phenol
  MW: 700-800   Melting Range C: 104-106
  CAS No. 68610-51-5   Physical Form: Powder, Pellets

Anox PP 18:
  Octadecyl-3-(3',5'-di-t-butyl-4'-hydroxyphenyl) propionate
  MW: 531   Melting Range C: 49-53
  CAS No. 2082-79-3   Physical Form: Powder, Granular,
                              NDB, Liquid

Anox 20:
  Tetrakismethylene(3,5-di-t-butyl-4-Hidroxyhydrocinnamate)
methane   Melting Range C: 110-125
  MW: 1178   Physical Form: Powder, Granular,
  CAS No. 6683-19-8   NDB, Amorphous

Great Lakes Chemical Corp.: Antioxidants (Continued):

Phenols (Continued):
Anox 29:
    2,2'-Ethylidenebis (4,6-di-tert-butylphenol)
    MW: 439   Melting Range C: 162-166
    CAS No. 35958-30-6   Physical Form: Powder

Anox IC-14:
    Tris (3,5-di-t-butyl-4-hydroxybenzyl) isocyanurate
    MW: 784   Melting Range C: 217-219
    CAS No. 27676-62-6   Physical Form: Powder, Granular

Anox 70:
    2,2' thiodiethyl bis-(3,5-di-tert-butyl-4-hydroxyphenyl)
    propionate
    MW: 642   Melting Range C: 63-73
    CAS No. 41484-35-9   Physical Form: Powder, Granular

Lowinox CA 22:
    1,1,3-Tris-(2'-methyl-4'-hydroxy-5'-tert-butyl-phenyl)-
    butane
    MW: 545   Melting Range C: 182-184
    CAS No. 1843-03-4   Physical Form: Powder

Lowinox WSP:
    2,2'-Methylene-bis-6-(1-methyl-cyclohexyl)-para-cresol
    MW: 421   Melting Range C: 132-137
    CAS No. 77-62-3   Physical Form: Powder

Lowinox HD 98:
    N,N'-hexamethylene bis (3,5-di-tert-butyl-4-hydroxy-hydro-
    cinnamamide)
    MW: 637   Melting Range C: 156-162
    CAS No. 23128-74-7   Physical Form: Powder, NDB

Lowinox GP 45:
    Triethyleneglycol-bis-(3-tert-butyl-4-hydroxy-5-methylphenyl)
    propionate   Melting Range C: 76-79
    MW: 587   Physical Form: Powder, Granular,
    CAS No. 36443-68-2   NDB

Anox BF:
    Benzenepropanionic acid, 3,5-bis(1,1-dimethylethyl)-4-
    hydroxy-C13,15-branched and linear alkyl esters
    MW: 485   Physical Form: Liquid
    CAS No. 171090-93-0

Great Lakes Chemical Corp.: Antioxidants (Continued):

Phosphites:
Alkanox 240:
  Tris (2,4-di-tert-butyl-phenyl) phosphite
  MW: 646   Melting Range C: 180-186
  CAS No. 31570-04-4   Physical Form: Powder, Granular,
                              NDB

Alkanox 240-3T:
  Tris (2,4-di-tert-butyl-phenyl) phosphite plus Distearyl-3,3-
  thiodipropionate (3% on phosphite)
  Melting Range: 63-186   Physical Form: Powder, Granular, NDB

Alkanox P-24:
  Bis (2,4-di-tert-butyl-phenyl) pentaerythritol-diphosphite
  MW: 604   Melting Range C: 160-175
  CAS No. 26741-53-7   Physical Form: Powder

Alkanox 24-44:
  Tetrakis (2,4-di-tert-butyl-phenyl) 4,4'-biphenylene-
  diphosphonite
  MW: 1035   Melting Range C: 75-95
  CAS No. 119345-01-6   Physical Form: Powder, Flakes
  CAS No. 38613-77-3

Alkanox TNPP:
  Tris (p-nonylphenyl)-phosphite
  MW: 688   Physical Form: Liquid
  CAS No. 26523-78-4

Thioesters:
Lowinox DLTDP:
  Di-lauryl-3,3'-thio-dipropionate
  MW: 515   Melting Range C: 38-41
  CAS No. 123-28-4   Physical Form: Flakes

Lowinox DSTDP:
  Di-stearyl-3,3'-thio-dipropionate
  MW: 683   Melting Range C: 63-67
  CAS No. 693-36-7   Physical Form: Flakes

Metal Deactivator:
Lowinox MD 24:
  1,2-BIS (3,5-di-tert-butyl-4 hydroxyhydrocinnamoyl) hydrazine
  MW: 553   Melting Range C: 221-232
  CAS No. 32687-78-8   Physical Form: Powder

Mayzo: Antioxidants:

BNX 1000:
   A synergistic liquid antioxidant system
   Formula: Proprietary
Characteristics:
   1. Non-staining and non-discoloring on light or heat aging.
   2. Because of its low volatility, the heat stability and
      antioxidation actions are excellent.
   3. Good extraction resistance by hot water and detergents.
   4. Essentially odorless.
   5. Provides long term stability for water base and hot melt
      adhesives.
   6. Provides better antioxidant properties than pigment
      dispersion systems of antioxidants.
   7. Contains an inhibitor for the catalytic oxidation incurred
      by metals.
   8. The synergistic characteristics of the system are achieved
      by:
      * incorporation of the antioxidants in a liquid polymer
        system which provides for slow release;
      * reduction of the antioxidants to the molecular level
        providing maximum antioxidant performance;
      * the liquid polymer base which ensures broad and effective
        dispersion in solvent and water base systems.

BNX (BENNOX) 1010/1010G:
   Tetrakis[Methylene-3(3',5'-di-tert-butyl-4-hydroxyphenyl)
propionate]methane
   CAS Number: 6683-19-8
Characteristics:
   1. Non-staining and non-discoloring on light or heat aging
   2. Because of its low volatility, the heat stability and
      antioxidation action is excellent.
   3. Good extraction resistance by hot water and detergents.
   4. Is odorless and tasteless.
   5. Has low oral and dermal toxicity.
   6. FDA cleared for use in adhesives and tackifiers.
   7. Provides long term stability in hot melt adhesives.

Mayzo: Antioxidants and Thermal Stabilizers:

BNX 1035:
Common Name: Thiodiethylene bis(3,5-di-tert-butyl-4-hydroxy)
hydrocinnamate
CAS Number: 41484-35-9
Empirical Formula: C38H58O6S
Uses:
  BNX 1035 is a high molecular weight hindered phenolic antioxidant. It provides excellent processing and end use stability
to a broad range of polymers.
  BNX 1035 is a particularly effective antioxidant for polyethylene, styrenic polymers, polypropylene and elastomers such
as EPDM and SBR and for carboxylated SBR latex, polybutadiene
rubber and polyisoprene rubber.
Characteristics:
  1. Non-staining and non-discoloring.
  2. Because of its low volatility, the heat stability and
     antioxidation actions are excellent.
  3. Thermally stable.
  4. Is odorless and tasteless.
  5. Has low oral and dermal toxicity.
  6. FDA cleared for use in adhesives, petroleum alicylic
     hydrocarbon resins, and various rubber articles.
  7. Especially useful in the stabilization of low density
     polyethylene for wire and cable coatings.

BNX 1076/1076G:
Stearyl-3-(3',5'-di-tert-butyl-4-hydroxyphenyl) propionate
CAS Number: 2082-79-3
Empirical Formula: C35H62O3
Uses:
  BNX 1076 is a high molecular weight hindered phenolic anti-
oxidant which effectively inhibits oxidation and thermal degradation of many organic and polymeric materials.
  BNX 1076 is an excellent antioxidant and stabilizer for
polyolefins (polyethylene, polypropylene, polymethylpentane,
etc.), polystyrene, ABS resin, methacrylic resin, PVC resin,
and EPDM.
  BNX 1076 is also effective for polyester resins, rubbers,
latex, varnish, adhesives, urethane and petroleum products.
Characteristics:
  1. Non-staining and non-discoloring on light or heat aging.
  2. Because of its low volatility, the heat stability and
     antioxidation action is excellent.
  3. Good extraction resistance by hot water and detergents.
  4. Thermally stable, odorless and tasteless.
  5. Easily incorporated by melt compounding techniques due to
     its low melting point.
  6. Retards light-induced deterioration of polymers.
  7. Has low oral and dermal toxicity.

Mayzo: BENEFOS 1680 Phosphite Type Antioxidant:

Chemical Name: Tris(2,4-di-tert-butylphenyl)phosphite
Chemical Family: Phosphite
CAS Number: 31570-04-4
Empirical Formula: C42H63O3P
Typical Physical Properties:
Form: White crystalline powder
Melting Range: 183-187C
Molecular Weight: 647
Specific Gravity (20C): 1.03 g/m3
Vapor Pressure (20C): 1 x 10 -18 mm Hg
Percent Volatiles: <0.5% maximum (2hr/105C)
Decomposition Temperature: >300C
Uses:

Benefos 1680 is a phosphite type antioxidant of low volatility and extremely high resistance to hydrolysis. It provides excellent antioxidant process stability to organic polymers, resistance to discoloration and adds long-term protection against thermo-oxidative degradation after processing.

Benefos 1680 is a secondary antioxidant. It is particularly useful in polyolefins and olefin-copolymers such as polyethylene, polypropylene, polybutene and ethylene vinyl acetate copolymers as well as polycarbonate and polyamids. Other applications include use in linear polyesters, high impact polystyrene, ABS, SAN, adhesives, natural and synthetic tackifier resins, elastomers such as BR, IR, and other organic substrates.

Neville Chemical Co.: NEVASTAIN 21 Nonstaining Antioxidant:

Nevastain 21 Nonstaining Antioxidant is a hindered phenolic compound which is produced by alkylating phenols with selected olefinic monomers.

Physical Properties:

	Typical Properties	Specifications
Color Gardner (ASTM D-1544):	5	7 Max.
Refractive Index @ 25C: (N9.12)	1.6006	1.5988-1.6024
Flash Point F (COC) (ASTM D-92):	365	----
Brookfield Viscosity @ 25C, cps: (N4.12)	5000+-2400	----
Form:	Liquid	

FDA Status:

Nevastain 21 Nonstaining Antioxidant is an approved substance as defined by the following United States Food and Drug Administration regulations:

175.105   Adhesives
177.2600   Rubber Articles Intended for Repeated Use

The formulator must comply with all other requirements of the FDA regulations, including conditions of use and extractive tolerances of the total compound or formula.

TSCA Status:

Nevile Chemical Co. certifies that this product has been registered in accordance with the rules and regulations of the Toxic Substances Control Act.

Raschiq Corp.: RALOX Antioxidants:

Ralox BHT Food Grade:
Ralox BHT Food Grade is a colorless, crystalline antioxidant and belongs to the group of non-discoloring, sterically hindered phenols.
Ralox BHT Food Grade is mainly used for the stabilization of articles made of polymers which come into contact with food-stuff (or which are used in the field of drinking water), polyols, polyurethanes, adhesives and hot melts for a possible contact with foodstuff, odoriferous substances and/or perfumes, foodstuff and printing colors. There is BGA and FDA approval for Ralox BHT food grade.

Chemical Structure:
    Empirical Formula: C15H24O
    Molecular weight: 220.4 g/mol
    CAS-No.: 128-37-0
    Registation: EINECS, TSCA, MITI
    HS-Code: 2907 19 90
Specification:
    Aspect: Colorless crystals
    Purity (GC): min. 99.5%
    Solidification point: min. 69.2C
    Colour (melt): max. 10 HZ
    Sulfate ash: max. 0.002%
    Arsenic content: max. 2 mg/kg
    Heavy metal content as lead: max. 10 mg/kg

Ralox LC:
Ralox LC is a light to cream-colored antioxidant belonging to the group of non-discoloring sterically hindered phenols. Since Ralox LC has an oligomer structure, it provides excellent protection against aging combined with low thermal volatility and extractability. Also, Ralox LC will not show any signs of discoloring, even when exposed to the light, a feature which is quite unlike any ortho-linked bisphenols.
Ralox LC is used primarily for stabilizing elastomers based on natural and synthetic rubbers, as well as ABS, hot melts and lubricants. Ralox LC is BGA and FDA approved.

Chemical Structure:
    Chemical description: Butylated reaction product of p-cresol
                          and dicyclopentadiene
    CAS No.: 68610-51-5
    Registration: EINECS, TSCA, MITI
    HS-Code: 2907 29 90
Specification:
    Appearance: Light to cream-colored powder or light yellowish
                to brown flakes
    Melting point: min. 105C
    Sulfate ash: max. 0.5% of weight

Raschig Corp.: RALOX Antioxidants (Continued):

Ralox TMQ:
Ralox TMQ is a light yellowish to brown antioxidant from the group of coloring, sterically hindered amines.
Ralox TMQ is a highly effective agent that prevents heat-induced aging and also provides minor protection against ozone and fatigue crack failure. Its major field of application is with black elastomers based on SBR, NR, IR, BR and NBR, as well as latices. A highly effective agent, Ralox TMQ is also added to polyethylene that has been cross-linked with peroxide.

Chemical Structure:
CAS No.: 26780-96-1
Registrations: EINECS (polymer), TSCA, MITI
HS-Code: 3812 30 80
Chemical description: 1,2-Dihydro-2,2,4-trimethyl chino-line, polymer

Specification:
Ralox TMQ-G:
Appearance: light brown to brown pastilles
Softening point (ring/ball): 86-100C
Sulfate ash: max. 0.4% of weight
Ralox TMQ-T:
Appearance: light yellowish to brown powder or light brown to brown pastilles
Melting point (capillary): min. 83C
Sulfate ash: max. 0.5% of weight
Ralox TMQ-H:
Appearance: light yellow powder
Melting point (capillary): min. 100C
Ash: Max. 0.5% of weight

Application:
Ralox TMQ is a highly effective agent that prevents heat-induced aging with black vulcanizates based on SBR, NR, IR, BR and NBR, as well as thermoplastic elastomers and their respective blends. Ralox TMQ also provides fairly good protection against rubber toxins (heavy metals). Regular doses of Ralox TMQ may vary between 0.5 and 1.5 phr. When combined with other antioxidants, 0.5-0.7 phr may be sufficient.

Raschig Corp.: RALOX Antioxidants (Continued):

Ralox 46:
Ralox 46 is a colorless to slightly cream-colored antioxidant belonging to the group of non-discoloring, sterically hindered phenols.
Ralox 46 is used primarily for natural and synthetic rubbers (SBR, BR, IR, NBR) and latices. Due to its effectiveness, Ralox 46 is also used as a stabilizing agent in polyacetates, ABS and mineral oil products that are exposed to high temperatures. Ralox 46 is FDA and BGA approved.

Chemical Structure:
   Empirical Formula: C23H32O2
   Molecular Weight: 340.5 g/mol
   CAS-No.: 119-47-1
   Registration: EINECS, TSCA, MITI
   HS-Code: 2907 29 90
Specification:
   Appearance: white to cream-colored powder
   Melting point: min. 125.0C
   Ash: max. 0.1% of weight
Application:
   Ralox 46 is a tried and tested antioxidant for natural and synthetic latices which by long-term experience has shown to provide optimum safety with respect to stabilization.

Ralox 926:
Ralox 926 is an antioxidant belonging to the group of non-discoloring, sterically hindered phenols. Due to its higher molecular weight, thermal volatility is lower in Ralox 926 than in BHT, making the product highly suitable for many forms of application, especially at high temperatures. As Ralox 926 is a fluid, exact doses are easy to apply.

Chemical Structure:
   CAS-No.: 4306-88-1
   Registration: EINECS, TSCA
   HS-Code: 2907 19 90
   Chemical Description: Synergistic mixture of butylated
                        p-alkylphenols
Specification:
   Appearance: clear, amber-colored fluid
   Water (Karl Fischer): max. 0.05% of weight
Application:
   Ralox 926 features excellent aging protection, and it can be applied in many different fields. Major fields of application are polyolefins, paraffin waxes, latices, polyols, benzenes and lubricants based on synthetic mineral oils or natural vegetable oils.

3V Inc.: ALVINOX Antioxidants:

Alvinox FB Antioxidant:
Chemical and Physical Characteristics:
  Appearance: white powder
  CAS #: 27676-62-6
  EINECS: 248-597-9
  Molecular Weight: 784.1
  Chemical Formula: C48H69N3O6
  Melting Point (C): 218-223
  Bulk Density (g/cc): 0.35-0.40
Properties:
  Alvinox FB is highly efficient primary antioxidant for poly-
mers. Its chemical structure includes three sterically hindered
phenolic groups. It functions mainly as "radical scavenger"; it
reacts efficiently with peroxy radicals, thus slowing down the
chain reaction process, which facilitates thermal oxidation.
  Alvinox FB either alone or combined with other antioxidants
(commercialized under the trade name Alvipack) is highly
efficient in a wide range of polymers such as:
* polyolefins and olefin copolymers (PP, HDPE, LDPE, LLDPE, EVA)
* polyamides   * polycarbonates
* styrenics (PS, ABS, IPS)   * adhesives
* elastomers   * latex

Alvinox P Antioxidant:
Chemical and Physical Characteristics:
  Appearance: white powder
  CAS #: 31570-04-4
  EINECS: 250-709-6
  Molecular Weight: 647
  Chemical Formula: C42H63O3P
  Melting Point (C): 183-186
  Bulk Density (g/cc): 1.2-1.6
Properties:
  Alvinox P is utilized to protect polymers from thermal
oxidation particularly during processing. It acts essentially
as a decomposer of hydroperoxides. Optimum performance is
obtained when the product is used in combination with antioxidants (hindered phenols and/or thioethers) which additionally
protect the polymer against long-term heat degradation.
Alvinox P combined with the previous antioxidants is highly
efficient in a wide range of polymers. 3V commercialized
pre-formed blends of Alvinox P and primary antioxidants under
the trade name of Alvipack. Their use is suitable in:
* polyolefins and olefin copolymers (PP, HDPE, LDPE, LLDPE, EVA)
* polyamides   * polycarbonates
* styrenics (PS, ABS, IPS)   * adhesives
* elastomers

3V Inc.: ALVINOX Antioxidants (Continued):

Alvinox 100 Antioxidant:
   C54H78O3
   Molecular weight: 775.2
   CAS number: 1709-70-2
   EINECS number: 2169710
Chemical and Physical Characteristics:
   Appearance: white powder
   Melting point: 240-245C
   Bulk density: approx. 0.47 g/cm3
Properties:
   Alvinox 100 is a highly efficient primary antioxidant for
polymers. Its chemical structure includes three sterically
hindered phenolic groups. It functions mainly as "radical
scavenger"; it reacts efficiently with peroxy radicals, thus
slowing down the chain reaction process which faclitates thermal
oxidation.
Alvinox 100 offers the following performance advantages:
   * High processing and long-term heat stabilization;
   * Low volatility in the range of temperatures normally used
     for processing and service life of polymers;
   * High compatibility with a wide range of polymers;
   * High compatibility and synergism with a broad range of
     additives used in polymer formulations (in particular with
     secondary antioxidants);
   * Very low extractability when in polymers in contact with
     water
   * Very good dielectric properties.
Applications:
   Alvinox 100 is particularly recommended to protect polyolefins
such as:
   * Homopolymer polypropylene (PP)
   * Low density polyethylene (LDPE)
   * High density polyethylene (HDPE)
   * Copolymers (EP, EVA)
Alvinox 100 can also be used in technopolymers like:
   * polyesters
   * polyamides
   * thermoplastic polyurethanes
and in:
   * adhesives and elastomers

3V Inc.: ALVIPACK Antioxidant Blends:

Alvipack 122 Antioxidant Blend:
Composition (nominal): Blend of Alvinox FB and Alvinox P and
            DSTDP (1:2:2)
Chemical and Physical Characteristics: Appearance: white powder
Properties:
    Alvipack 122 is a highly efficient antioxidant blend of:
Alvinox FB (a primary phenolic antioxidant) and
Alvinox P (an organophosphite of low volatility and high hydro-
lytic stability) and
DSTDP (a thioether)
    It functions both as a radical scavenger, reacting with
peroxy radicals, and as a decomposer of hydroperoxides formed
during thermal oxidation of polymers.
Alvipack 122 offers the following performance advantages:
* Good processing and long term heat stabilization
* Low volatility and high thermal stability in the range of
  temperatures used during processing and service life of
  polymers.
* High hydrolytic stability.
* Excellent color stability during processing and end use of
  polymers.
* High compatibility and synergism with a wide range of
  plastic additives used in the formulation of polymers.
* High resistance to gas fading caused by interaction with
  other additives as well as urban pollution.

Alvipack 12 Antioxidant Blend:
    Chemical Composition: 33% Alvinox FB
    (on weight basis)   67% Alvinox P
    Components: Alvinox FB: Alvinox P:
        CAS #: 27676-62-6   CAS #: 31570-04-4
        EINECS: 248-597-9   EINECS: 250-709-6
Properties:
    The specific composition of the blend offers a good balance
of processing and long-term heat stability. The two antioxidants
work in order to fully protect the plastic from the processing to
the in use conditions.
Applications:
    Alvipack 12 offers the following performance advantages:
* high processing stabilization due to the massive presence
  of a phosphite antioxidant
* long term stabilization due to the presence of a phenolic
  antioxidant (for severe long term storage condition ask
  for Alvipack 11)
* low volatility and high thermal stability in the range of
  temperatures employed during processing and service life
  of polymers
* high hydrolytic stability
* good color stability during processing and end use of
  polymers
* high compatibility and synergism with a wide range of
  additives
* high resistance to gas fading caused by pollution agents

3V Inc.: ALVIPACK Antioxidant Blends (Continued):

Alvipack 11 Antioxidant Blend:
Chemical and Physical Characteristics:
  Chemical Composition: 50% Alvinox FB
    (on weight basis) 50% Alvinox P
  Appearance: white powder
Components:
  Alvinox FB: CAS #: 27676-62-6
             EINECS: 248-597-9
  Alvinox P: CAS #: 31570-04-4
             EINECS: 250-709-6
Properties:
  The specific composition of the blend offers a good balance
of processing and long-term heat stability. The two antioxidants
work in order to fully protect the plastic from the processing
to the in use conditions.
  The combination of the two products improves with a synergistic action the polymer stabilization. Optimum performance is
obtained when both antioxidants are used together such as in
Alvipack 11. This is highly efficient in a wide range of polymers
such as:
  * polyolefins and olefin copolymers (PP, HDPE, LDPE, LLDPE,
    EVA)
  * polyamides   * polycarbonates
  * polyesters   * styrenics (PS, ABS, IPS)
  * adhesives   * elastomers
Alvipack 11 offers the following performance advantages:
  * long term heat stabilization, thanks to the massive
    presence of phenolic antioxidant
  * moderate process stabilization due to the presence of a
    phosphite in the antioxidant mixture (for severe process
    conditions ask for Alvipack 12)
  * low volatility and high thermal stability in the range of
    temperatures employed during processing and service life
    of polymers
  * high hydrolytic stability
  * good color stability during processing and end use of
    polymers
  * high compatibility and synergism with a wide range of
    additives used in the formulation of polymers (antioxidants,
    light stabilizers, etc.)
  * high resistance to gas fading caused by urban and industrial
    pollution agents as anitrogen oxides

Uniroyal Chemical Co., Inc.: NAUGARD Antioxidants:

Naugard A Antioxidant:
   Naugard A is a cost effective amine based antioxidant finding
use in polyamides and carbon black filled olefin formulations
for use in geomembranes, wire and cable jacketing, and irrigation
piping.
Product Features:
   *Low dust flake   *Synergistic
   *Cost effective   *Easy dispersion
Chemical Composition: Acetone Diphenyl amine condensation product
Typical Properties:
   Appearance: Greenish-Tan flake or powder
   Melt Point Range: 85C-95C
   Color-Gardner: 14
   Specific Gravity: 1.15
   Flash Point (TCC): 179C

Naugard BHT Antioxidant:
   Naugard BHT is a cost effective, general purpose antioxidant
that is used in a wide variety of materials: polyethylene, poly-
styrene, polypropylene, hot-melt adhesives and coatings for
food packages, for example.
Product Features:
   *FDA regulated   *Economical
   *Direct food application
Chemical Structure: 2,6-di-tert-butyl hydroxytoluene
Typical Properties:
   Appearance: White crystals
   Melt Point: 69C
   Color-APHA: 10 (50 max)
   Specific Gravity @ 25C: 1.05
   Flash Point (TOC): 127C
   Molecular Weight: 220

Naugard HM11 Antioxidant:
   Naugard HM11 is a granular blend of hindered phenolic and
diphenylamine antioxidants. This blend provides synergistic
short and long term thermal protection against oxidation in
polypropylene.
Product Features:
   *Synergistic stabilization   *Low volatility
   *Non-dusting granular form   *FDA regulated
   *Polymer and rosin ester compatible
Chemical Structure: Blend of tetrakis [Methylene (3,5-di-t-
   butyl-4-hydroxyhydrocinnamate)] methane and 4,4'-bis (alpha,
   alpha-dimethylbenzyl) diphenylamine
Typical Properties:
   Appearance: White granules
   Color-APHA: 25
   Specific Gravity @ 25C: 1.16
   Flash Point (COC): 296C
   Molecular Weights: 1178/406

Uniroyal Chemical Co., Inc.: NAUGARD Antioxidants (Continued):

Naugard HM22 Antioxidant:
    Naugard HM22 is a granular blend of hindered phenolic and
diphenylamine antioxidants. This blend provides synergistic
short and long term thermal protection against oxidation in
polypropylene. It also exhibits excellent color and viscosity
stability, as well as gelation and skinning resistance in SIS,
SBS, EVA, PE, and polyamide hot melt adhesives.
Product Features:
    *Synergistic stabilization   *Low volatility
    *Non-dusting granular form   *FDA regulated
    *Polymer and rosin ester compatible
Chemical Structure: Blend of Octadecyl 3,5-di-tert-butyl-4-
    hydroxyhydrocinnamate and 4,4'-bis (alpha, alpha-dimethyl-
    benzyl) diphenylamine.
Typical Properties:
    Appearance: White granules
    Color-APHA: 25
    Specific Gravity @ 25C: 1.09
    Flash Point (COC): 277C
    Molecular Weights: 531/406

Naugard P Antioxidant:
    Naugard P is a cost effective secondary liquid phosphite anti-
oxidant that functions as a peroxide decomposer and as a processing stabilizer. When used in conjunction with hindered phenolic
and thiodipropionate antioxidants, a synergism is seen. In addition to improved antioxidant activity of these combinations,
Naugard P also inhibits discoloration of the polymer caused by
some phenolic antioxidants. Polymer applications include polyolefins, styrenics and a variety of other polymers.
Product Features:
    *FDA regulated   *Economical
    *Non-discoloring   *Low volatility
    *Liquid physical form
Chemical Structure: Tris (monononylphenyl) phosphite
Typical Properties:
    Appearance: Clear viscous liquid
    Viscosity @ 25C: 5,200 cps
    Color-Gardner: 2
    Specific Gravity @ 25C: 0.978
    Flash Point (TOC): 221C
    Pour Point: -7C
    Molecular Weight: 688

Uniroyal Chemical Co., Inc.: NAUGARD Antioxidants (Continued):

Naugard PHR Antioxidant:
   Naugard PHR is a hydrolytically resistant version of Naugard
P. It is a cost effective secondary liquid phosphite antioxidant
that functions as a peroxide decomposer and as a processing
stabilizer. When used in conjunction with hindered phenolic and
thidipropionate antixidants, a synergism is seen. In addition to
improved antioxidant activity of these combinations, Naugard PHR
also inhibits discoloration of the polymer caused by some phen-
olic antioxidants. Naugard PHR has applications in a wide variety
of polymers including polyolefins, styrenics and other various
polymers.
Product Features:
   *FDA regulated   *Economical
   *Non-discoloring   *Low volatility
   *Liquid physical form
Chemical Structure: Tris (monononylphenyl) phosphite
Typical Properties:
   Appearance: Clear viscous liquid
   Viscosity @ 25C: 5,200 cps
   Color-Gardner: 2
   Specific Gravity @ 25C: 0.978
   Flash Point (TOC): 221C
   Pour Point: -7C
   Molecular Weight: 688

Naugard SA Antioxidant:
   Naugard SA is a high performance, high temperature antioxidant
which finds use as a co-stabilizer in carbon black filled olefin
formulations for applications such as geomembranes, wire and
cable jacketing, and irrigation piping. Strong positive syner-
gism with other antioxidants makes Naugard SA a prime candidate
as an antioxidant choice for these applications.
Product Features:
   *High performance   *Synergistic
   *Easy dispersion
Chemical Structure: p-(p-toluene-sulfonylamido)-diphenylamine
Typical Properties:
   Appearance: Gray Powder
   Melt Point: 146C
   Specific Gravity: 1.32
   Flash Point (TCC): 271C
   Molecular Weight: 338

Uniroyal Chemical Co., Inc.: NAUGARD Antioxidants (Continued):

Naugard SP Antioxidant:
Naugard SP is a liquid phenolic antioxidant. Naugard SP is a
very economical alternative to conventional antioxidants.
Product Features:
 *FDA regulated   *Economical
 *Liquid physical form
Chemical Structure: Styrenated Phenol (Mixture of mono-, di- and
                        tri)
Typical Properties:
 Appearance: Transparent viscous liquid
 Viscosity 25C, cps: 2500
 Color-Gardner: 3
 Specific Gravity, 24C: 1.08
 Flash Point (COC): 184C
 Average Molecular Weight: 276
Naugard Super Q Antioxidant:
 Naugard Super Q is an effective processing and long term heat
stabilizer for a variety of application areas in LDPE, LLDPE,
HDPE, and ethylene-propylene compolymers. It is particularly
efficient in carbon black-filled polymer systems such as wire
and cable and geomembranes.
Product Features:
 *Non-dusting pastilles   *Low volatility
 *Low acute toxicity   *Cost effective heat stabilizer
Chemical Structure: Polymerized 1,2-dihydro-2,2,4-trimethyl-
                        quinoline
Typical Properties:
 Physical Form: Amber to Brown Pastille
 Softening Point: 128C
 Color-Gardner (G-136-A): 11
 Specific Gravity @ 25C: 1.08
 Flash Point (COC): 274C-282C
 Molecular Weight: 874
Naugard XL-1 Antioxidant:
 Naugard XL-1 is a unique antioxidant which incorporates a
metal deactivation function in the same product. This product
may be used where there is interference from metallic ions such
as from residual polymer catalyst, inorganic pigments or mineral
filled polymers. It is also unique in that it is FDA regulated
for use in most polymers. This material can be used in a wide
variety of polyolefinic and polystyrenic resins.
Product Features:
 *FDA regulated   *Non-discoloring
 *Dual functional antiox-   *Synergist with other AO's
  idant/metal deactivator
Chemical Structure: 2,2'-oxamido bis-[ethyl 3-(3,5-di-tert-
                        butyl-4-hydroxyphenyl)propionate]
Typical Properties:
 Appearance: White to Off-White Powder Flash Point(TOC): 260C
 Melting Point Range: 170C-180C   Molecular Weight: 697
 Color, Trans @ 425nm: 98.0
 Specific Gravity @ 20C: 1.12

Uniroyal Chemical Co., Inc.: NAUGARD Antioxidants (Continued):

Naugard 10 Antioxidant:
   Naugard 10 is a versatile antioxidant capable of giving
excellent processing stability and long term heat aging char-
acteristics. It is used in a variety of polymers where elevated
processing temperatures are encountered. It is particularly
effective in high density polyethylene and polypropylene.
Product Features:
   *Excellent long-term stability   *FDA regulated
   *Numerous physical forms   *Synergist with other AO's
Chemical Structure: Tetrakis [Methylene (3,5-di-t-butyl-4-
                     hydroxyhydrocinnamate)] methane
Typical Properties:
   Appearance: White to Off-White Powder (or alternate physical
               forms)
   Melting Point Range: 118C-127C
   Color, Trans @ 425nm: 95.0
   Specific Gravity @ 20C: 1.2
   Flash Point (TCC): 299C
   Molecular Weight: 1178

Naugard 76 Antioxidant:
   Naugard 76 is a cost effective antioxidant used in a wide
variety of polymers. It has low color generation charactreistics
and is used whenever color is of major importance. This product
effectively reduces degradation of most polymeric substances
and can be used in a wide variety of polymers such as linear
low density polyethylene and polystyrene.
Product Features:
   *Excellent color stability   *FDA regulated
   *Numerous physical forms   *Synergist with other AO's
Chemical Structure: Octadecyl 3,5-di-tert-butyl-4-hydroxyhydro-
                     cinnamate
Typical Properties:
   Appearance: White to Off-White Powder (or alternate physical
               forms)
   Melting Point Range: 51.5C-52.5C
   Color, Trans @ 425nm: 97.0
   Specific Gravity @ 20C: 1.02
   Flash Point (TCC): 273C
   Molecular Weight: 531

Uniroyal Chemical Co., Inc.: NAUGARD Antioxidants (Continued):

Naugard 88L:
Naugard 88L is a liquid hindered phenolic for use as an antioxidant in many applications. It is a low volatility, low viscosity antioxidant providing good protection against degradation
due to heat and oxygen. Performance of Naugard 88L is enhanced
when used in combination with amine or phosphite antioxidants.
Product Features:
    *Non-discoloring   *Synergistic
    *Liquid physical form   *Cost effective
    *Low viscosity   *Low volatility
Chemical Structure: 3,5-di-tert-butyl-4-hydroxyhydrocinnamic
                        acid, C7-C9 branched alkyl ester
Typical Properties:
    Appearance: Clear, Liquid Yellow Liquid
    Viscosity, cSt (40C): 105
    Color, Gardner: 1
    Specific Gravity @ 25C: 0.985
    Flash Point (PMCC): 148C
    Average Molecular Weight: 391

Naugard 431 Antioxidant:
Naugard 431 is a liquid phenolic antioxidant designed for
applications where accurate control of antioxidant concentrations
on a continuous basis is desirable or where optimum dispersion
of the antioxidant is needed. Naugard 431 shows good performance
in a wide range of polyolefins.
Product Features:
    *FDA regulated   *Lower volatility than BHT
    *Liquid physical form   *Economical
Chemical Structure: 2,6-Bis (∝-methylbenzyl)-4-methylphenol
Typical Properties:
    Appearance: Amber viscous to semicrystalline liquid
    Viscosity 38C, cps: 6200
    Color-Gardner: 2
    Specific Gravity @ 25C: 1.08
    Flash Point (COC): 199C
    Molecular Weight: 315

Naugard 438L:
Naugard 438L is a liquid nonylated diphenylamine antioxidant
finding use in carbon black pigmented styrenic (ABS) and poly-
ethylene formulations. Its liquid form allows for easy handling
and incorporation into various matrices.
Product Features:
    *Cost effective   *Liquid physical form
    *Synergistic
Chemical Structure: Major Component: Di-nonyl Diphenylamine
Typical Properties:
    Appearance: Dark brown, viscous liquid
    Kinematic Viscosity @ 40C, cSt: 650
    Color, Gardner: 8
    Nitrogen, wt%: 3.5   Flash Point (PMCC): 154C
    Specific Gravity, @ 25C/25C: 0.95 Average Molecular Wt: 397

Uniroyal Chemical Co., Inc.: NAUGARD Antioxidants (Continued):

Naugard 445 Antioxidant:
  Naugard 445 is a highly effective non-discoloring aromatic
amine type antioxidant finding utility as a thermal stabilizer
in a wide variety of applications, including polyolefins,
styrenics, polyols, hot melt adhesives, lubricants and poly-
maides. Excellent performance at processing temperatures, strong
synergy with other types of antioxidants such as phenolics and
phosphites, makes Naugard 445 an excellent choice as a component
of your thermal stabilizer package.
Product Features:
  *White powder   *Synergistic
  *Low dust granules available *High melting
Chemical Structure: 4,4'-Bis (α,α-dimethylbenzyl) diphenylamine
Typical Properties:
  Appearance: White powder or granules
  Melt Point Range: 98C-100C
  Color-APHA: 20
  Specific Gravity: 1.14
  Flash Point (TCC): 277C
  Molecular Weight: 406

Naugard 492 Phenolic/Phosphite:
  A new, low volatility antioxidant, Naugard 492 is a candidate
for many food-contact applications. Naugard 492, a phenolic
phosphite, not only offers low volatility, but is economical,
non-staining and non-discoloring. Applications include polyprop-
ylene, high and low density polyethylene, high impact polysty-
rene and ABS.
Features:
  *Low cost compared with BHT
  *Non-discoloring and non-staining
  *Low volatility
  *Acts as both radical scavenger and peroxide decomposer
  *Easily emulsifiable
  *FDA approved
Applications Include:
  *Polypropylene   *HDPE
  *LDPE   *HIPS
  *ABS   *PVC
  *Elastomers
Typical Properties:
  Chemical composition: Phenolic/phosphite
  Appearance: Light yellow, transparent liquid
  Flash point, Penske Marten: 315F (157C)
  Fire point, (TOC): 234F (112C)
  Specific Gravity @ 25C: 1.008
  Solubility: Completely soluble in acetone, Benzene, Toluene,
              Hexane and Cyclohexane. Partially soluble in
              Methanol. Insoluble in water.

Uniroyal Chemical Co., Inc.: NAUGARD Antioxidants (Continued):

Naugard 524:
Naugard 524 is a solid phosphite antioxidant with excellent
hydrolytic stability. Recommended for use in combination with
amine/phenolic antioxidants, Naugard 524 functions synergistically to retard oxidative degradation of most polymeric substances
during polymerization, processing, and in end-use applications.
For those polymers resistant to oxidation, Naugard 524 can be
used alone to maintain optimal color stability.
Product Features:
    *Excellent hydrolytic stability
    *Provides good long-term heat aging
    *High temperature processing stability
    *Non-discoloring
    *FDA regulated
    *Low oral and dermal toxicity
Chemical Structure: Tris (2,4-di-tertiary-butylphenyl) phosphite
Typical Properties:
    Appearance: White, free-flowing powder
    Melt Point Range: 180C-186C
    Color-APHA: 50
    Specific Gravity @ 20C: 1.03
    Flash Point (COC): 220C
    Molecular Weight: 647

Naugard 536 Antioxidant:
Naugard 536 is a viscous liquid phenolic antioxidant. Direct
metering of Naugard 536 allows for safe handling and accurate
additive addition. It was specifically designed to be used in
emulsion polymerization processes such as in ABS.
Product Features:
    *Easily emulsifiable   *Low volatility
    *Liquid physical form   *Economical
    *Imparts low odor and taste *FDA regulated
    *Non-discoloring
Chemical Structure: Major Component: 2,2'-Methylene bis
    (4-methyl-6-nonylphenol)
Typical Properties:
    Appearance: Clear, gold viscous liquid
    Viscosity @ 30C: 75 poise
    Color-Gardner: 6
    Specific Gravity @ 25C: 0.96
    Flash Point (TCC): 52C
    Molecular Weight: 481

Antioxidants 71

Uniroyal Chemical Co., Inc.: NAUGARD/WYTOX Antioxidants (Continued):

Naugard 635 Antioxidant:
Naugard 635 is a liquid styrenated diphenylamine for use as an antioxidant in wire and cable, adhesive, and elastomer applications. Naugard 635 can easily be incorporated into various matrices and can be synergistically combined with hindered phenolic antioxidants.

Product Features:
*FDA regulated   *Economical
*Liquid physical form
Chemical Structure: Styrenated diphenylamine, predominately para-substituted
Typical Properties:
Appearance: Clear, reddish liquid
Kin. Viscosity @ 40C, cSt: 760
Color-Gardner: 2
Specific Gravity @ 25C: 1.09
Flash Point, C: 237
Molecular Weight (Ave): 352

Wytox 312 Antioxidant:
Wytox 312 is a cost effective secondary liquid phosphite antioxidant that functions as a peroxide decomposer and as a processing stabilizer. When used in conjunction with hindered phenolic and thiodipropionate antioxidants, a synergism is seen. In addition to improved antioxidant activity of these combinations, Wytox 312 also inhibits discoloration of the polymer caused by some phenolic antioxidants. Polymer applications include polyolefins, styrenics and a variety of other polymers.

Product Features:
*FDA regulated   *Economical
*Non-discoloring   *Low volatility
*Liquid physical form
Chemical Structure: Tris (monononylphenyl) phosphite
Typical Properties:
Appearance: Clear viscous liquid
Viscosity @ 25C: 5,200 cps
Color-Gardner: 2
Specific Gravity @ 25C: 0.978
Flash Point (TOC): 221C
Pour Point: -7C
Molecular Weight: 688

Uniroyal Chemical Co., Inc.: POLYGARD Antioxidants:

Polygard Antioxidant:
Polygard is a cost effective secondary liquid phosphite antioxidant that functions as a peroxide decomposer and as a processing stabilizer. When used in conjunction with hindered phenolic and thiodipropionate antioxidants, a synergism is seen. In addition to improved antioxidant activity of these combinations, Polygard also inhibits discoloration of the polymer caused by some phenolic antioxidants. Polymer applications include polyolefins, styrenics and a variety of thermoplastics.

Product Features:
  *FDA regulated   *Economical
  *Non-discoloring   *Low volatility
  *Liquid physical form

Chemical Structure: Tris (mixed mono- and dinonylphenyl) phosphite

Typical Properties:
  Appearance: Clear viscous liquid
  Viscosity @ 25C: 5,200 cps
  Color-Gardner: 3
  Specific Gravity @ 25C: 0.99
  Flash Point (PMCC): 140.5C
  Pour Point: -7C
  Molecular Weights: 688/1066

Polygard HR Antioxidant:
Polygard HR is a cost effective secondary liquid phosphite antioxidant that functions as a peroxide decomposer and as a processing stabilizer. When used in conjunction with hindered phenolic and thiodipropionate antioxidants, a synergism is seen. In addition to improved antioxidant activity of these combinations, Polygard HR also inhibits discoloration of the polymer caused by some phenolic antioxidants. Polymer applications include polyolefins, styrenics and a variety of thermoplastics.

Product Features:
  *FDA regulated   *Economical
  *Non-discoloring   *Low volatility
  *Liquid physical form

Chemical Structure: Tris (mixed mono- and dinonylphenyl) phosphite

Typical Properties:
  Appearance: Clear viscous liquid
  Viscosity @ 25C: 5,200 cps
  Color-Gardner: 3
  Specific Gravity @ 25C: 0.99
  Flash Point (PMCC): 140.5C
  Pour Point: -7C
  Molecular Weights: 688/1066

Witco Corp.: Additives for Polyolefins: Recommended Uses: Antioxidants:

DLTDP (Thio):
FDA Sanctioned: Yes
HDPE & Polypropylene:
Antioxidant Extrusion/Antioxidant Film/Antioxidant General Purpose/Antioxidant Molding

DMTDP (Thio):
FDA Sanctioned: Yes
HDPE & Polypropylene:
Antioxidant Extrusion/Antioxidant Film/Antioxidant General Purpose/Antioxidant Molding

DSTDP (Thio):
FDA Sanctioned: Yes
Polypropylene:
Antioxidant Extrusion/Antioxidant Film/Antioxidant General Purpose/Antioxidant Molding

Mark 328 (Blend):
FDA Sanctioned: Yes
HDPE & Polypropylene:
Antioxidant Extrusion/Antioxidant Film/Antioxidant General Purpose/Antioxidant Molding

Mark 1589 (Blend):
FDA Sanctioned: Yes
HDPE & LDPE & LLDPE & UHMWPE:
Antioxidant Extrusion/Antioxidant Film/Antioxidant General Purpose/Antioxidant Molding
Polypropylene: Antioxidant All Purpose

Mark 1589B (Blend):
FDA Sanctioned: Yes
HDPE & LDPE & LLDPE:
Antioxidant Extrusion/Antioxidant Film/Antioxidant General Purpose/Antioxidant Molding
Polypropylene: Antioxidant All Purpose

Mark 5004 (Blend):
FDA Sanctioned: Yes
Polypropylene: Antioxidant

Mark 5111 (Blend):
FDA Sanctioned: Yes
HDPE & LDPE & LLDPE & UHMWPE:
Antioxidant Extrusion/Antioxidant General Purpose/Antioxidant Other/Antioxidant Molding
Polypropylene: Antioxidant All Purpose

Witco Corp.: Additives for Polyolefins: Recommended Uses:
Antioxidants (Continued):

Mark 5116 (Blend):
HDPE & LDPE & LLDPE:
Antioxidant Extrusion/Antioxidant Film/Antioxidant General
Purpose/Antioxidant Molding
Polypropylene: Antioxidant All Purpose

Mark 5118A (Blend):
HDPE & LDPE & LLDPE:
Antioxidant Extrusion/Antioxidant Film/Antioxidant General
Purpose/Antioxidant Molding
Polypropylene: Antioxidant All Purpose

Mark 5121 (Blend):
FDA Sanctioned: Yes
HDPE:
Antioxidant Extrusion/Antioxidant Film/Antioxidant General
Purpose/Antioxidant Molding
Polypropylene: Antioxidant All Purpose

Mark 260 (Phosphite):
HDPE & LDPE & LLDPE & Polypropylene & UHMWPE:
Antioxidant All Purpose

Mark 522 (Phosphite):
UHMWPE: Antioxidant General Purpose/Antioxidant Fiber

Mark 1178 (Phosphite):
FDA Sanctioned: Yes
EVA Modified PE/HDPE/LDPE/LLDPE/UHMWPE:
Antioxidant Extrusion/Antioxidant Film/Antioxidant General
Purpose/Antioxidant Molding
Polypropylene: Antioxidant All Purpose

Mark 1178B (Phosphite):
FDA Sanctioned: Yes
EVA Modified PE/HDPE/LDPE/LLDPE/UHMWPE:
Antioxidant Extrusion/Antioxidant Film/Antioxidant General
Purpose/Antioxidant Molding
Polypropylene: Antioxidant All Purpose

Mark 5060 (Phosphite):
FDA Sanctioned: Yes
EVA Modified PE/HDPE/LDPE/LLDPE/UHMWPE:
Antioxidant Extrusion/Antioxidant Film/Antioxidant General
Purpose/Antioxidant Molding
Polypropylene: Antioxidant All Purpose

**Witco Corp.: Additives for Polyolefins: Recommended Uses:
Antioxidants (Continued):**

Mark 5082 (Phosphite):
FDA Sanctioned: Yes
EVA Modified PE/UHMWPE:
   Antioxidant Extrusion/Antioxidant General Purpose/Antioxidant Molding/Antioxidant Other
HDPE/LDPE/LLDPE:
   Antioxidant Extrusion/Antioxidant Film/Antioxidant General Purpose/Antioxidant Molding
Polypropylene: Antioxidant All Purpose

Mark 2140 (Thio):
Polypropylene:
   Antioxidant Extrusion/Antioxidant Molding/Antioxidant General Purpose

Mark 5095 (Thio):
FDA Sanctioned: Yes
HDPE/Polypropylene:
   Antioxidant Extrusion/Antioxidant Film/Antioxidant General Purpose/Antioxidant Molding

Seenox 412S (Thio):
HDPE/Polypropylene:
   Antioxidant Extrusion/Antioxidant Film/Antioxidant General Purpose/Antioxidant Molding

Section IV
Anti-Static Agents

ACL Staticide: STATICIDE Topical Anti-Stats:

Staticide, an inexpensive, easy-to-use topical anti-static solution, is a highly-effective method for long-term static control. The family of solutions are widely used in a broad range of industries, as well as in the office environment.

3000 Staticide Concentrate:
Closest thing yet to "Static Control Insurance."
Staticide is the key ingredient in many ACL anti-static solutions. Diluted and mixed with water or various solvents, it eliminates a host of static control problems--especially in th electronics, textile and plastics industries. Staticide Concentrate is available to save freight, handling and storage costs while offering greater flexibility in dilution.

Ready-to-Use Dilutions
Heavy Duty Staticide for Porous Surfaces:
Ideal for eliminating static related problems such as:
*Static electricity discharge on carpeting & fabrics
*Jamming or slipping of materials during printing, packaging, or converting
*Ignition of combustible vapors, dust or solvents, causing fire or explosion
One gallon covers approximately 2,000 to 2,500 square feet and will last from weeks to months, depending on the application.

General Purpose Staticide for Non-Porous Surfaces:
Solves static related problems such as:
*The attraction of dirt, dust and bacteria to all environ-mental surfaces, plastic products and product packaging.
*Charge generation on surfaces of tote boxes and carriers used to process and store electronic components
*Data processing "glitches"...memory loss, data errors, paper jams
One gallon covers approximately 2,000 to 2,500 square feet despnding on the application.

Staticide Family Features:
*Can be applied by spraying, wiping, dipping or by transfer roller coating, gravure coating or flexographic printing
*Meets MIL-B-81705 and NFPA-56A electrostatic decay criteria when tested in accordance with Federal Test Standard 101B, Method 4046
*Proven effective in relative humidity below 15%
*Non-staining, completely biodegradable and safe to use

ACL Staticide: STATICIDE Topical Anti-Stats (Continued):

Staticide ESD Permanent Aerosol Spray Coating:

Stop static electricity permanently with new Staticide ESD clear aerosol coating

Permanent aerosol spray coating can be applied to plastics, painted metals, laminates, paper and other materials. Independent laboratory tests show that Staticide ESD meets the EOS/ESD association's standard test S11.11 for permanent electrostatic protection.

Staticide ESD works as a static electricity Problem Solver and may be applied to keyboards, computers, mice, oscilloscopes, multi-meters, shelving, workstations, mini-environments and wherever dangerous static electricity invades and threatens your electronic equipment.

Simply spray Staticide ESD on any surface to get instant permanent static protection (108 ohms @ 10 volts) even down to zero humidity. Staticide ESD meets Federal Standard 209E for zero particulation in cleanrooms. Staticide ESD meets California air quality standards and is exempt from V.O.C. reporting as an ESD aerosol product.

Anti-Static Foam Cleaner for Glass and Plastic by Staticide:

ACL/Staticide's Anti-static Foam Cleaner for Glass & Plastic is specially formulated for general external cleaning of computer hardware, glass and plastic materials/machines in the office and home. Leaves no film while keeping equipment clean and static free.

The special Foam Cleaning formulation provides a safe, quick and unique way to clean computer equipment, fax machines, copy machines, TV screens, glass and plastic. Foam properties reduce the potential of moisture damage that can be caused by liquid cleaners. Anti-static properties help repel dirt and dust, keeping equipment cleaner longer. ACL/Staticide's Anti-static Foam Cleaner for Glass & Plastic decreases the number of cleanings needed on your home and office equipment, saving the user time and money.

Akzo Nobel Chemicals Inc.: ARMOSTAT Antistatic Additives:

Armostat 310:
    Bis(2-hydroxyethyl)tallowamine
    CAS No.: 61791-44-4
    TSCA status: On Inventory
    Permanent internal antistatic agent for use in high, low and
linear low density polyethylene and polypropylene
Characteristics:
    Tertiary amine: 97% min
    Moisture: 0.5% max
    Specific Gravity @ 25C: 0.916
    Melting point: 32C (90F)
    Appearance: liquid to paste
    Color, Gardner (1963): 1 max

Armostat 410:
    Bis(2-hydroxyethyl)cocoamine
    CAS No.: 61791-31-9
    TSCA status: On Inventory
    Permanent internal antistatic agent for use in polyethylene,
polypropylene, polystyrene, ABS, HIPS and SAN resins.
Characteristics:
    Tertiary amine: 97% min
    Moisture: 0.5% max
    Specific gravity @ 25C: 0.874
    Melting point: 12C (54F)
    Appearance @ 25C: clear liquid
    Color, Gardner (1963): 2 max

Armostat 710:
    Bis(2-hydroxyethyl)oleylamine
    CAS No.: 25307-17-9
    TSCA status: On Inventory
    Permanent internal antistatic agent for use in high and low
density polyethylene
Characteristics:
    Tertiary amine: 97% min
    Moisture: 0.5% max
    Specific gravity @ 20C: 0.916
    Melting point: -17C (1F)
    Appearance: clear liquid
    Color, Gardner (1963): 2 max

Akzo Nobel Chemicals Inc.: ARMOSTAT Antistatic Additives (Continued):

Armostat 1800:
   Bis(2-hydroxyethyl)stearylamine
   CAS No.: 10213-78-2
   TSCA status: On Inventory
   Permanent internal antistatic agent for use in high and low density polyethylene, polypropylene and biaxially oriented polypropylene (BOPP) film.
   Characteristics:
   Tertiary amine: 97% min
   Moisture: 0.5% max
   Specific gravity @ 60C: 0.876
   Melting point: 52C (126F)
   Appearance: White solid
   Color, Gardner (1963): 2 max

Armostat 2000 flakes:
   Lauric diethanol amide
   Specification:
   Appearance: off-white flakes
   Melting point: 40C min
   Water content: 0.8% max
   Color, 50C: 250 Pt-Co max
Permanent antistat for polyolefins

Armostat 3002:
   Sodium alkane sulphonate
   Specification:
   Appearance: off-white flakes
   Assay: min. 93.0%
   Sodium chloride: max. 6.0%
   Water: max. 2.0%
Permanent antistat for thermoplastics

Akzo Nobel Chemicals Inc.: ARMOSTAT Antistatic Additive Concentrates:

Armostat 350:
　　Composition: Armostat 310, 50% in LDPE
　　Form: Pellets
　　Bulk Density, untapped/tapped (g/mL): 0.48/0.51

Armostat 375:
　　Composition: Armostat 310, 75% in HDPE
　　Form: Pellets
　　Bulk Density, untapped/tapped (g/mL): 0.53/0.56

Armostat 450:
　　Composition: Armostat 410, 50% in PS
　　Form: Pellets
　　Bulk Density, untapped/tapped (g/mL): 0.58/0.61

Armostat 475:
　　Composition: Armostat 410, 75% in PP
　　Form: Pellets
　　Bulk Density, untapped/tapped (g/mL): 0.49/0.51

Armostat 550:
　　Composition: Armostat 410, 50% in SAN
　　Form: Pellets
　　Bulk Density, untapped/tapped (g/mL): 0.59/0.63

Amstat Industries, Inc.: Anti-static Chemicals:

Staticide:
Is an easy-to-use topical antistatic solution widely used in
a broad range of industrial and office applications. It is easily
applied to walls, ceiling, floors, carpets, fixtures, work
surfaces, tools and equipment, conveyor belts, and other materials.
Use Heavy Duty Staticide for porous, absorbent, or high friction surfaces. On a hard surface, one gallon of Heavy Duty Staticide will cover several thousand square feet and last from weeks
to months, depending upon the application.
Use General Purpose Staticide for non-porous or low friction
surfaces or clear high gloss materials.

Static-Blok FDA-3:
Anti-Static Solution is a long lasting industrial strength
anti-stat suitable for use on food contact surfaces. It may be
wiped, dipped, or sprayed as need dictates, and is wear and rub-
off resistant. Static-Blok FDA-3 is excellent on PVC, polystyrene, nylon and other plastics.
Static-Blok FDA-3 Anti-Static Solution is an aqueous solution
of one of a group of surface active materials regulated by the
FDA under CFR 178.3400 for use in the manufacture of articles
that contact food. There is no limitation in this regulation of
the kind of food allowed to contact an article containing Static-
Blok FDA-3, or on the kind of material (plastic, etc.) of which
the article is made, or on the level of Static-Blok FDA-3 except
for the general "good manufacturing practice" requirements to
use no more than necessary and that "the quantity that may become
a component of food shall not be intended to, nor in fact, to
accomplish any physical or technical effect in the food itself."

Form: Liquid   Solubility in Isopropanol: 100%
Color: Clear, colorless   pH (approximate): 7
Odor: Practically odorless   Stability: Stable
Chemical family: Polyether
Solubility in water: 100%

Surfaces properly treated with Static-Blok FDA-3 Anti-Static
Solution meet or exceed standards for electrostatic decay times
as set forth by NFPA 56A and MIL B-81705B, in accordance with
Federal Test Standard 101C, Method 4046.
Static-Blok FDA-3 Anti-Static Solution is available in ready-
to-use solutions and in 10:1 concentrate. 10:1 concentrate may
be diluted with water and/or isopropyl. A gallon of 10:1 concen-
tate yields 11 diluted gallons.

BASF Corp.: LAROSTAT Antistats:

Larostat 377DPG Antistat:
   Larostat 377 DPG is a mixture of n-alkyl dimethyl ethyl
ammonium ethyl sulfates in dipropylene glycol. Excellent heat
stability and low odor content are two of the attributes of
the product. It is an easily handled liquid and is an esp-
ecially effective antistat. Use levels of 1% to 2% are typical
to achieve a static half life of approximately one second.
Typical Physical Properties:
   Active Quaternary: 78%-82%
   Dipropylene Glycol: 20%
   pH, 10% Solution: 6-7
   Viscosity: 1000 cps
   Color, Gardner: 5
Specifications:
   Acid Value: 8 Max
   Appearance @ 25C: Fluid
   Activities, % (375): 76 Min

Larostat 902A & 902AS:
   Larostat 902A is a custom made antistat based on unique
alkanolamine chemistry. This chemistry was developed for use
in polycarbonates for electronic applications. Larostat 902A
is non-reactive with polycarbonate and less corrosive than
amines.
   Larostat 902A is very effective in polyolefins. In certain
formulations, it meets Mil Spec B-8105C performance. The
Larostat 902A is a liquid product, but is offered in a powdered
form as a 60% active version on silica, Larostat 902AS.

Typical Properties:
902A:
   Boiling Point, F: >200
   Solubility in Water @ 25C: Soluble
   Specific Gravity @ 25C: 0.981
   Appearance @ 25C: Clear Liquid
   Odor: Mild
   Flashpoint, PMCC, F: >200
902AS:
   Boiling Point, F: >350
   Solubility in Water @ 25C: Partial
   Appearance @ 25C: White Pwd.
   Odor: Mild
   Flashpoint, PMCC, F: >200

BASF Corp.: LAROSTAT Antistatic Additives:

BASF Specialty Products offers an extensive line of antistatic additives for a variety of applications. Through novel, patented chemistry, BASF has developed a complete line of internal as well as topical additives. These products include sulfate quats, amides, phosphate esters and ethoxylated amines.

264A:
  H2O Soluble: Y
  Conc.: 35
  Internal: X
  Topical: X
  Form: Liquid
264A Anhy.:
  H2O Soluble: Y
  Conc.: 99
  Internal: X
  Topical: X
  Form: Waxy Solid
  Notes: Solvent soluble
377 DPG:
  H2O Soluble: Y
  Conc.: 80
  Internal: X
  Form: Liquid
519:
  H2O Soluble: Dispersible
  Conc.: 60
  Internal: X
  Form: Powder
902A:
  H2O Soluble: Y
  Conc.: 99
  Internal: X
  Topical: X
  Form: Liquid
  Notes: Non-amine
902AS:
  H2O Soluble: Dispersible
  Conc.: 60
  Internal: X
  Form: Powder
  Notes: Stable to 230C on silica
1084:
  H2O Soluble: Y
  Conc.: 97
  Topical: X
  Form: Liquid
  Notes: Lubricant

FPE S:
  H2O Soluble: Dispersible
  Conc.: 60
  Internal: X
  Form: Powder
  Notes: FPE on silica
HTS 905:
  H2O Soluble: Y
  Conc.: 90
  Internal: X
  Form: Liquid
  Notes: Stable to 250C
HTS 905S:
  H2O Soluble: Dispersible
  Conc.: 60
  Internal: X
  Form: Powder
  Notes: HTS 905 on silica

BASF Corp.: LAROSTAT FPE-S Food Grade Antistat:

Larostat FPE-S is an effective antistat for polyolefins that can be used in food packaging and housewares applications. It can be compounded into plastic films without imparting a greasy feel to the surface and is more antistatically active and less corrosive than ethoxylated amine based antistats. Larostat FPE-S exhibits good water resistance when internally incorporated in polyethylene films. It can also be applied as an effective topical antistat and is water soluble at room temperature. Larostat FPE-S offers the following advantages:

*Meets FDA Title 21, paragraph 178.3130 for an antistatic agent at levels not to exceed 0.5 weight percent of molded/ extruded polyethylene containers intended for contact with honey, chocolate, liquid sweeteners, condiments, flavor extracts and liquid flavor concentrates, grated cheese, light and heavy cream, yogurt, and any food type that is a dry solid with the surface containing no free fat or oil.

For use in polypropylene as an antistat agent, not to exceed 0.2 weight percent of the types of films described in 177.1520 and 176.170(c).

FDA regulations can be very complicated. Product use determinations involve both chemistry and specific applications. The ultimate decision on compliance must be made by the user with full understanding of the application.

*Safe and effective for housewares applications: polyethylene and or polypropylene straws and polystyrene cups.

*Does not affect polyethylene film quality or clarity at use concentrations.

*Can be used to meet Military Specification B-81705B in non-FDA packaging.

*Resistant to prolonged shower exposure.

*Effective at 12% relative humidity.

*Less corrosive than ethoxylated amine antistats.

Larostat FPE-S is a 60% active powder with the following properties:
Typical Physical Properties:
    Boiling Point, F: >200
    Solubility in Water @ 25C: Dispersible
    Appearance @ 25C: White Powder
    Odor: Mild Ester
    Bulk Density, Kg/L: 0.5
    Flash Point, PMCC, F: >200
Specifications:
    Residue on Ignition, %: 38.0-42.0

BASF Corp.: LAROSTAT HTS 905 Antistat:

Larostat HTS 905 Antistat:
   Most antistats degrade at temperatures of 200-300F. Larostat
HTS 905 was specifically developed for use in high temperature
applications, >300F. Larostat HTS 905 has a much higher thermal
stability (stable to 250C) than standard antistats and can
provide long term, internal static protection.
Applications:
   Larostat HTS 905 is used in Engineered Thermoplastics, Flex-
ible PVC, Polypropylene, Textiles and Dust control.
Typical Physical Properties:
   Boiling Point, F: 220
   Solubility in Water @ 25C: Soluble
   Specific Gravity @ 25C: 1.073
   Volatiles, %, By Volume: 11
   Appearance @ 25C: Amber Liquid
   Odor: Bland
   Flash Point, PMCC, F: >200
   Concentration: 90
Specifications:
   Color, Gardner (1963): 1.0 Max.
   Acid Value: 2.0 Max.
   Amine Value: 1.0 Max.
   pH, 10% Aq: 6.0 to 8.5
   Solids, % @ 105C, 1.5 Hrs.: 86.0 to 92.0

Chemax, Inc.: CHEMSTAT Antistats:

Liquid: Chemstat	Use Level % by Weight PE	PP	PVC	PS/ABS
106G/90			0.50-1.0	0.10-3.0
122	0.10-0.40	0.15-0.80		1.0- 4.0
172T	0.10-0.40			
182	0.10-0.40			
1913/50			1.0-1.5	2.0-3.0
P-300	1.5-3.0			
P-400	1.5-3.0			
SE-5	1.0-1.5			
SE-20	1.0-1.5			

106G/90:
    Main Function/Classification: External/Internal/Cationic
    Chemical Composition: Quaternary Compound
    FDA: No
122:
    Main Function/Classification: Internal/Cationic
    Chemical Composition: Ethoxylated coco amine
    FDA: Yes
172T:
    Main Function/Classification: Internal/Cationic
    Chemical Composition: Ethoxylated oleyl amine
    FDA: Yes
182:
    Main Function/Classification: Internal/Cationic
    Chemical Composition: Ethoxylated tallow amine
    FDA: Yes
1913/50:
    Main Function/Classification: External/Internal/Cationic
    Chemical Composition: Quaternary Compound
    FDA: No
P-300:
    Main Function/Classification: Internal/Nonionic
    Chemical Composition: Polyethylene Glycol
    FDA: Yes
P-400:
    Main Function/Classification: Internal/Nonionic
    Chemical Composition: Polyethylene Glycol
    FDA: Yes
SE-5:
    Main Function/Classification: External/Nonionic
    Chemical Composition: Ethoxylated sorbitan ester
    FDA: No
SE-20:
    Main Function/Classification: External/Nonionic
    Chemical Composition: Ethoxylated sorbitan ester
    FDA: Yes

Chemax, Inc.: CHEMSTAT Antistats (Continued):

Solid/Pellet:
Chemstat:
273-E:
  Use Level % by Weight: PP: 0.15-0.80
  Chemical Composition: Ethoxylated stearyl amine
  FDA: Yes

AC-100:
  Use Level % by Weight: PE: 0.50-1.00%
  Chemical Composition: Proprietary blend
  FDA: Yes

AC-1000:
  Use Level % by Weight: PE: 2.5-8.0
  Chemical Composition: Lauric diethanolamide
  FDA: Yes

AC-201:
  Use Level % by Weight: PE: 0.50-1.0
  Chemical Composition: Proprietary blend
  FDA: Yes

AC-2000:
  Use Level % by Weight: PP: 3.0-8.0
  Chemical Composition: Lauric diethanolamide
  FDA: Yes

AC-101:
  Use Level % by Weight: PS/ABS: 3.0-5.0
  Chemical Composition: Sodium alkyl sulfonate
  FDA: Yes

LD-100:
  Use Level % by Weight: PE: 0.15-1.2//PP: 0.25-1.2
  Chemical Composition: Lauric Diethanolamide
  FDA: Yes

LX-1000:
  Use Level % by Weight: PU: 0.50-2.0
  Chemical Composition: Proprietary blend
  FDA: Yes

PS-101:
  Use Level % by Weight: PVC: 1.0-1.5//PC: 1.5-2.5//PET: 1.0-2.0
  Chemical Composition: Sodium alkyl sulfonate
  FDA: Yes

Chemax, Inc.: CHEMSTAT Antistats (Continued):

Powder:
106G/60DC:
    Use Level % by Weight: PS/ABS: 0.25-3.0
    Chemical Composition: Quaternary compound
    FDA: No

122/60DC:
    Use Level % by Weight: PE: 0.15-0.60//PP: 0.25-1.4//
                            PS/ABS: 1.5-4.0
    Chemical Composition: Ethoxylated coco amine
    FDA: Yes

182/67DC:
    Use Level % by Weight: PE: 0.15-0.60
    Chemical Composition: Ethoxylated tallow amine
    FDA: Yes

192/NCP:
    Use Level % by Weight: PP: 0.20-0.90//PS/ABS: 1.5-4.0
    Chemical Composition: Ethoxylated stearyl amine
    FDA: Yes

6000/50DC:
    Use Level % by Weight: PP: 0.30-0.80
    Chemical Composition: Proprietary blend
    FDA: Yes

LD-100/60DC:
    Use Level % by Weight: PE: 0.25-1.4//PP: 0.25-1.4
    Chemical Composition: Lauric diethanolamide
    FDA: Yes

PS-101/PWDR:
    Use Level % by Weight: PVC: 1.0-1.5//PS/ABS: 1.5-2.5
                            PC: 2.5-3.5//PET: 1.0-2.0
    Chemical Composition: Sodium alkyl sulfonate
    FDA: Yes

Cytec Industries Inc.: CYASTAT Antistatic Agents:

Cyastat antistatic agents from Cytec are quaternary ammonium compounds available in solid or solution form for use as internal or external treatment of plastic and other materials.

Solids:
LS:
Chemical Formula: C19H42O5N2S
CAS Number: 010595-49-0
A quaternary ammonium compound (QAC) for internal application in polymers and elastomers with some thermal stability limitations.

Solutions:
SN:
Chemical Formula: C25H53O5N3
CAS Number: 002764-13-8
A 50 percent active QAC antistatic agent solution which prevents the accumulation of static charge on a variety of polymer surfaces.

SP:
Chemical Formula: C25H55O6N2P
CAS Number: 3758-54-1
A 35 percent active QAC antistatic agent solution suitable for systems that are alkaline due to the buffering effect of the dihydrogen phosphate anion.

609:
Chemical Formula: C21H47O8NS
CAS Number: 018602-17-0
A 50 percent active QAC antistatic agent solution with outstanding performance suitable for surface treatment of polymers used in electronic parts packaging.

Cytec Industries Inc.: CYASTAT Antistatic Agents:

Cyastat AA-609
    antistatic agent is noted for its excellent antistatic activity. Additionally, its good heat stability minimizes undesirable color generation associated with many antistatic agents when applied internally. It is particularly effective in polyvinyl chloride and polyester compositions and can be applied as a surface treatment or incorporated internally by conventional formulation procedures.
    Cyastat AA-609, topically applied, allows the electrostatic spray painting of non-metallic substrates, such as those used for automotive OEM cars and trucks.

Cyastat AA-SN
    antistatic agent prevents accumulation of static charge on a wide variety of polymers. It can be incorporated internally into many polymers ambient cured or baked at low temperatures. It can also be applied as a surface treatment where heat stability is a problem. Both methods are effective, although incorporation generally results in greater retention of antistatic properties.
    Cyastat AA-SN used in surface coatings and other materials serves a dual function, acting as a dispersant as well as an antistatic agent.

Cyastat AA-LS
    antistatic agent is useful in a wide variety of coatings. Its powder form particularly lends itself to internal incorporation. As a solution, it can also be applied externally.

Cyastat AA-SP
    antistatic agent imparts surface-active antistatic properties to non-metallic substrates. This cationic liquid product can also serve as an emulsifier, settling, dispersing and rewetting agent.
    The buffering effect of the dihydrogen phosphate anion makes Cyastat AA-SP more suitable for alkaline systems, such as non-acid cured coating systems. At concentration loadings of 1% in acid cured coating systems, cure interference may occur.

Eastern Color & Chemical Co.: ECCOSTAT ASP Anti-Static Agent:

An organo-phosphonate based anti-static agent

Properties:
Appearance: Clear water-white to light straw colored viscous
liquid
Active Content: 100%
Solubility: Dispersible in plastics, monomer, solvents and
water
Density: 10.7 lbs/gal.

General Comments:
Eccostat ASP is a new anti-static agent designed for optimum control of static in the processing of plastics, rubber and polymers.
Eccostat ASP is suggested for application in plastic and rubber compounding to control static. It is also recommended for use in extruded fibers and fibers of polyester, nylon and olefins.
Eccostat ASP will not cause any discoloration nor impair the tensile strengths of fibers.
Eccostat ASP is also useful in solvent-based paints, adhesives and other non-aqueous systems.

Application:
The use of 3 to 8% on total weight of Eccostat ASP is suggested for optimum results.
The exact amount depends upon polymer type and antistatic requirements.
Apply in compounding or as a masterbatch.

JacksonLea: LEA Antistat Anti-Static Solution

Lea Antistat when applied to plastic and non-metallic surfaces, neutralizes ststic charge build-up--a common source for attracting lint and dirt in buffing operations, storage or other handling. Lea Antistat leaves a transparent colorless coating on the surface.

Two methods are used in applying Lea Antistat anti-static solution to surfaces:

Transferral by buffing
Hand wiping

In production buffing, a clean, soft wipe-off buff is used for the final buffing operation. The face of the wipe-off buff is coated with a spray of Lea Antistat and the surfaces are then gone over completely with the buff. A thin coating is transferred from the buff to the surface in this way.

Lea Antistat can easily be sprayed, using a simple plunger type atomizer such as is used with window cleaner solutions.

The second method of applying Lea Antistat is to spread it sparingly by hand with a soft cloth or spray it directly on the plastic surface, allow it to dry for a few minutes and then wipe off the excess with a clean, soft cloth.

Lonza Inc.: GLYCOSTAT and GLYCOLUBE Antistatic Agents:

Glycolube 140:
  Polyol Ester
  Major Application: Thermoplastic antistat for PP and PE

Glycolube AFA-1:
  Ester Blend
  Major Application: Antistat for PVC, antifog for PP, PE, PVC,
  PS

Glycostat:
  Polyol Ester
  Major Application: Antistat for PP, PE

Product Class Benefits:
  The Glycolube series of polymer additives offers unique performance properties which increase the efficiency of plastics
processing and final polymer compounding. Lonza offers a broad
range of lubricants, antistats and antifog agents for all major
polymer classes. Many of the Glycolube products have FDA
clearance as indirect food additives in food packaging applications.

Glycostat Plastic Antistatic Agent:
Physical Properties:
  Acid Value: 2 Max.
  Saponification Value: 160-175
  Color, Gardner 1963: 3 Max.
Suggested Applications:
  Glycostat was developed for use as a nonionic thermoplastic
resin antistatic agent. It is suggested for use in:
  Polyolefins: HDPE, LDPE, Polypropylene
  PVC
  Polystyrene: crystal and impact grades
  Glycostat is permitted by the FDA for food contact applica-
tions.

Merix Chemical Co.: MERIX Anti-Static #79 Concentrate:

For Hard Surfaces:
* Plastics & paper
* Solar conductors
* Computers & electronics

Function:
Stops static and removes electrostatic charges and related
dust and dirt attraction on all types of hard surfaced plastics
such as acrylics, polystyrenes, PVC, styrenes, vinyl, thermo-
plastics, and thermosets. Stops static on films and paper.
Assists solar heat-transfer.
Acts as conductive coating in electronic and computer applications: #79 diluted 1:1 reported to give readings of 20 to
100 megohms per square on plastics, 100/up megohms per square
on glass surfaces.
Appearance:
Clear, neutral, invisible after water dilution when dried.
Flashpoint: None. Totally non-flammable
Application Method:
Surface destaticizing by dip, wipe, mop, spray, brush or
calendar methods. Drying time: A few seconds when wiped on,
longer for other methods.
Internal as an additive to water-soluble paints and inks
and to coating materials and adhesives in the manufacture of
static-free coated papers.
Merix Anti-Static #79 Concentrate gives enduring protection
against electrostatic-caused dust and dirt attraction on
plastics, films, and paper from one application. New potency
allows dilution with water up to twenty parts, cutting destaticizing costs by at least 1/20th! New additives assure perfect
run-off, fast-drying, and spot-free high lustre finish in
dipping, spraying, brushing, or wipe-on methods. All types of
plastics, when so treated on either one side or both, will stay
clean and static-free.

#79-OL Concentrate (OdorLess):
For soft surfaces
* Carpets
* Textiles
* Polyethylenes

Merix Anti-Static PARABOLIX 100:
Each gallon Parabolix 100 diluted 1:100 can clean and add
sheen up to approximately 200,000 sq.ft. of parabolic luminaires.

Merix Anti-Static #79 Special:
Low sodium/low calcium concentrate
Hard surfaces and plastics used in highly sensitive applications such as micro-conductors, delicate instruments, highly-
sensitive microelectronic products.

Patco Additives Division: PATIONIC Antistatic Agents:

Pationic 901:
    Pationic 901 is a distilled (95% minimum, monoester) glycerol monostearate derived from fully hydrogenated vegetable oil and produced to polymer additive heat stability requirements.
Typical Applications:
    Antistatic Agent-Polyolefins--Effective at low levels. Its efficiency is attributed to the high concentration of highly functional monoester.
    Mold Release Agent-Polypropylene--Effective at 0.2-0.5%.
    Internal Lubricant-PVC
    Melt Flow Enhancement
    Colorant Dispersion/Color Concentrate Distribution

Pationic 902:
    Pationic 902 is a distilled (95% minimum, monoester) glycerol monostearate derived from fully hydrogenated animal fat and produced to polymer additive heat stability requirements.
Typical Applications:
    Antistatic Agent-Polyolefins--Effective at low levels. Its efficiency is attributed to the high concentration of highly functional monoester.
    Mold Release Agent-Polypropylene--Effective at 0.2-0.5%.
    Internal Lubricant-PVC
    Melt Flow Enhancement
    Colorant Dispersion/Color Concentrate Dispersion

Pationic 909 Specialty Polymer Additive:
    Pationic 909 is a distilled (95%, minimum monoester) glycerol mono stearate derived from fully hydrogenated vegetable oil and produced to polymer additive heat stability requirements.
Designed Uses:
    Designed for use in polymer systems and compounds requiring dry blending or surface coating of the additive directly onto resin (or compound), pellets, powder, or beads. Its fine particle size allows for simplified dry blending of Pationic 909 with other polymer additives and ingredients.
Typical Applications:
    EPS Processing Aid--Pationic 909 is surface coated onto polystyrene beads prior to the pre-expansion process. The additive provides lubrication minimizing lumping and caking, and it acts as a processing antistatic agent.
    Colorant Dispersant and Lubricant
    Color Concentrate Distribution

Patco Additives Division: PATIONIC Antistatic Agents (Continued):

Pationic 919 Specialty Polymer Additive:
Pationic 919 is derived from fully hydrogenated vegetable oil; it is an edible food product known chemically as glycerol tri stearate (GTS).

Typical Current Applications:
EPS Processing Aid--Pationic 919 is surface coated onto polystyrene beads prior to the pre-expansion process. The additive provides lubrication minimizing lumping and caking.
Color Dispersant & Lubricant
Color Concentrate Distribution

Pationic 1042:
Pationic 1042 is a glycerol mono stearate product with a 42%, minimum, alpha-monoester content. It is derived from natural fats and oils. A Kosher grade (1042K) is available.

Typical Current Applications:
Pationic 1042 is used as an additive to polyolefins, styrenics, PVC, TPE's and TPO's. The product can be used as an antistatic agent, an internally compounded mold release additive, a flow modifier, and an internal lubricant. It is specifically designed for use in the plastics industry by polymer producers, compounders, and concentrate manufacturers.

Pationic 1042K:
Pationic 1042K is a Kosher glycerol mono stearate product with a 42%, minimum, alpha-monoester content. It is derived from natural fats and oils. A non-Kosher grade (1042) is available.

Typical Current Applications:
Pationic 1042K is used as an additive to polyolefins, styrenics, PVC, TPE's and TPO's. The product can be used as an antistatic agent, an internally compounded mold release additive, a flow modifier, and an internal lubricant. It is specially designed for use in the plastics industry by polymer producers, compounders, and concentrate manufacturers.

Pationic 1052:
Pationic 1052 is a glycerol mono stearate product with a 52%, minimum, alpha-monoester content. It is derived from natural fats and oils. A Kosher grade (1052K) is available.

Typical Current Applications:
Pationic 1052 is used as an additive to polyolefins, styrenics, PVC, TPE's and TPO's. The product can be used as an antistatic agent, an internally compounded mold release additive, a flow modifier, and an internal lubricant. It is specifically designed for use in the plastics industry by polymer producers, compounders, and concentrate manufacturers.

Patco Additives Division: PATIONIC Antistatic Agents (Continued):

Pationic 1052K:
Pationic 1052K is a glycerol mono stearate product with a 52%, minimum, alpha-monoester content. It is derived from natural fats and oils. A non-Kosher grade (1052) is available.
Typical Current Applications:
Pationic 1052K is used as an additive to polyolefins, styrenics, PVC, TPE's and TPO's. The product can be used as an antistatic agent, an internally compounded mold release additive, a flow modifier, and an internal lubricant. It is specifically designed for use in the plastics industry by polymer producers, compounders, and concentrate manufacturers.

Pationic AS10:
Pationic AS10 is a self-emulsifiable glycerol ester antistatic agent derived from natural oils.
Primary Application:
Pationic AS10 is applied to PVC resin following polymerization, prior to drying, to inhibit the generation of static charge.
Primary Benefits:
Treating PVC with Pationic AS10 will maintain the bulk density of the resin when it is exposed to various sources that induce static charge, such as transfer lines, packaging equipment and storage silos. Resin processors will observe improved flowability and handling characteristics of treated resin in blending equipment and feed hoppers which also induce static charge. In addition, Pationic AS10 has been shown to improve resin heat stability.

Pationic AS22:
Pationic AS22 is a glycerol ester antistatic agent derived from natural oils.
Primary Application:
Pationic AS22 is applied to PVC resin to inhibit the generation of static charge. The product is designed for either addition directly into the slurry tank following polymerization or direct spraying onto the PVC "wetcake".
Primary Benefits:
Treating PVC with Pationic AS22 will maintain the bulk density of the resin when it is exposed to various sources that induce static charge, such as transfer lines, packaging equipment and storage silos. Resin processors will observe improved flowability and handling characteristics of treated resin in blending equipment and feed hoppers, which also induce static charge.

Patco Additives Division: PATIONIC Antistatic Agents (Continued):

Pationic AS33:
    Pationic AS33 is a glycerol ester antistatic agent derived
from natural triglycerides.
Primary Application:
    Pationic AS33 is applied to PVC resin to inhibit the genera-
tion of static charge. The product is designed for either addition directly into the slurry tank following polymerization or
direct spraying onto the PVC "wetcake".
Primary Benefits:
    Treating PVC with Pationic AS33 will maintain the bulk density
of the resin when it is exposed to various sources that induce
static charge, such as transfer lines, packaging equipment and
storage silos. Resin processors will observe improved flowability
and handling characteristics of treated resin in blending equipment and feed hoppers, which also induce static charge.

Pationic AS38 Antistatic Agent for PVC Resin:
    Pationic AS38 is a low viscosity, high activity, proprietary
ester antistatic agent derived from natural oils.
Primary Application:
    Pationic AS38 is applied to PVC resin to inhibit the genera-
tion of static charge. The product is designed for either addition directly into the slurry tank following polymerization
or direct spraying onto the PVC "wetcake". The affinity for the
PVC resin exhibited by Pationic AS38 minimizes loss of the
active antistat to the water phase during centrifuging. The
low viscosity contributes to uniform distribution when spraying
"neat" onto the "wetcake". The need for a water emulsion of the
antistat is eliminated and no additional water is introduced
into the system prior to drying. A water emulsifiable system
is available, if preferred.
Primary Benefits:
    Treating PVC with Pationic AS38 will maintain the bulk dens-
ity of the resin when it is exposed to various sources that
induce static charge, such as transfer lines, packaging equipment and storage silos. Resin processors will observe improved
flowability and handling characteristics of treated resin in
blending equipment and feed hoppers, which will also induce
static charge. Pationic AS38 enhances the heat stability of the
PVC resin.

Pationic AS40:
    Pationic AS40 is a self-emulsifiable glycerol ester antistatic
agent derived from natural oils.
Primary Application:
    Pationic AS40 is applied to PVC resin following polymeriz-
ation, prior to drying, to inhibit the generation of static
charge. It can be applied neat or as an aqueous emulsion.
Primary Benefits:
    Treating PVC with Pationic AS40 will maintain the bulk dens-
ity of the resin when it is exposed to various sources that
induce static charge, such as transfer lines, packaging equipment and storage silos.

3V Inc.: ANTISTATICO KN Antistatic Agent:

C25H53N2O2-NO3
1-Propanaminium, N-(2-hydroxyethyl)-N,N-dimethyl-3-(1-oxooctadecyl) amino]-,nitrate (salt) in lower alcohol-water solution
    Molecular weight: 475.8
    CAS number: 2764-13-8
    EINECS number: 2204325

Chemical and Physical Characteristics:
    Appearance: clear slightly yellow liquid
    Boiling point: approx. 100C
    Density: 0.95-0.97 g/cm3 at 20C
    Dry matter: 50+-1%
    pH: 6+-1
    Solubility:
        Antistatico KN can be mixed with water, acetone, alcohols
and other polar solvents of low molecular weight. Solutions of
different concentrations can be prepared with other solvents,
if required, by heating.
Properties:
    Antistatico KN operates both by dissipating electrostatic
charges and by preventing their formation. The surface resistivity can be reduced by 10 3 to 10 5 order of magnitude. When
compounded, it exerts a lubricating effect on the compound.
    It is compatible with non-ionic and with cationic agents.
It is compatible with anionic agents only in determined propor-
tion or when the reaction product is soluble.
    Antistatico KN has a good stability to acids and to alkalis,
however, its behaviour in these conditions should be tested
case by case.

Applications:
    Antistatico KN can be employed either externally or inter-
nally; in fact it imparts protection in both instances, to:
    * PVC (rigid)
    * Acrylics
    * Polystyrene
    * ABS
    * Polyolefins
External application:
    Plastic materials can be treated by applying a solution of
Antistatico KN by spraying or dipping, or by rubbing them with
a drenched cloth. In the first two cases 3V recommends using a
1 to 10% solution in reference to the dry matter, while in the
last case a 10 to 30% solution always in reference to the dry
matter, is preferable.
Internal Application:
    For internal applications 3V recommends incorporating a
quantity variable from 0.5% to 4% of Antistatico KN, depending
on the type of polymer used.
    Antistatico KN is well suited for internal use in molding
grades of polymeric materials.

Tomah Products Inc.: E-Series Ethoxylated Amines: Anti-Static Agent Use:

E-14-2:
Description: iso-(2-hydroxyethyl) isodecyloxypropylamine
E-14-5:
Description: poly (5) oxyethylene isodecyloxypropylamine
E-17-2:
Description: bis-(2-hydroxyethyl) isotridecyloxypropylamine
E-17-5:
Description: poly (5) oxyethylene isotridecyloxypropylamine
E-19-2:
Description: bis-(2-hydroxyethyl) linear C12-15 alkyloxypropylamine

Other Ethoxylated Fatty Amines:
E-18-2:
Description: bis-(2-hydroxyethyl) octadecylamine (& 5, 15 mole adduct)
E-S-2:
Description: bis-(2-hydroxyethyl) soya amine (& 15 mole adduct)
E-T-2:
Description: bis-(2-hydroxyethyl) tallow amine (& 5, 15 mole adduct

Typical Properties:

E-14-2:
Combining Wt.: 310
Min. Amine Value: 175
E-14-5:
Combining Wt.: 445
Min. Amine Value: 123
E-17-2:
Combining Wt.: 345
Min. Amine Value: 155
E-17-5:
Combining Wt.: 485
Min. Amine Value: 112
E-19-2:
Combining Wt.: 350
Min. Amine Value: 150
E-18-2:
Combining Wt.: 360
Min. Amine Value: 153
E-18-5:
Combining Wt.: 490
Min. Amine Value: 112

E-18-15:
Combining Wt.: 450
Min. Amine Value: 58
E-S-2:
Combining Wt.: 374
Min. Amine Value: 150
E-S-15:
Combining Wt.: 925
Min. Amine Value: 57
E-T-2:
Combining Wt.: 360
Min. Amine Value: 156
E-T-5:
Combining Wt.: 485
Min. Amine Value: 112
E-T-15:
Combining Wt.: 925
Min. Amine Value: 59

Tomah Products, Inc.: Q-Series Quaternary Amines: Anti-static Agent Use:

Q-14-2:
74% active isodecyloxypropyl bis-(2-hydroxyethyl) methyl ammonium chloride (Available in propylene glycol as Q-14-2 PG).

Q-17-2:
74% active isotridecyloxypropyl bis-(2-hydroxyethyl) methyl ammonium chloride. (Available in propylene glycol as Q-17-2 PG).

Q-17-5:
74% active isotridecyloxypropyl poly (5) oxyethylene methyl ammonium chloride (Available in propylene glycol).

Q-18-2:
50% active octadecyl bis-(2-hydroxyethyl) methyl ammonium chloride (15 mole ethylene oxide adduct also available).

Q-S:
50% & 80% active mono soya methyl ammonium chloride.

Q-DT:
50% active tallow diamine diquaternary. (Available at 70% active in hexylene glycol as Q-DT HG).

Q-C-15:
100% active coco poly (15) oxyethylene methyl ammonium chloride

Q-ST-50:
50% active trimethyl stearyl quaternary

Typical Properties:

Q-14-2:
pH, 5% solution: 6-9
Combining Wt: 370

Q-17-2:
pH, 5% solution: 6-9
Combining Wt.: 410

Q-17-5:
pH, 5% solution: 6-9
Combining Wt.: 535

Q-18-2:
pH, 5% solution: 6-9
Combining Wt.: 415

Q-S:
pH, 5% solution: 6-9
Combining Wt.: 410

Q-S-80:
pH, 5% solution: 6-9
Combining Wt.: 410

Q-D-T:
pH, 5% solution: 5-9
Combining Wt.: 525

Q-DT-HG:
pH, 5% solution: 5-8
Combining Wt.: 525

Q-C-15:
pH, 5% solution: 6-9
Combining Wt.: 915

Q-ST-50:
pH, 5% solution: 5-9
Combining Wt.: 348

Section V
Antibacterials/Fungicides/ Mildewcides

Ferro Corp.: MICRO-CHEK Antimicrobials:

Micro-Chek is an EPA registered mildewcide for PVC, polyurethane and other polymer compositions susceptible to attack by microorganisms. With proper use, Micro-Chek will help prevent microbiological attack on the product surface that can cause the loss of aesthetic appearance, mildew odors, embrittlement and premature product failure. Micro-Chek is a 4.0% solution of the active ingredient 2-n-octyl-4-isothiazoline-3-one in several different carriers. The isothiazolinone is non-leaching which enables Micro-Chek to effectively provide protection in the most stressful of environments: roofing membranes, automotive trim, awnings, pond liners, marine upholstery, shower curtains and outdoor furniture. Micro-Chek can also provide superior performance in less severe applications such as wall coverings and appliance gasketing.

Use Data:
Micro-Chek is recommended to be used between 1.0 to 2.5% of the total formulation weight. This usually corresponds to 2.0-4.5 parts per hundred resin depending upon the specific product application.

Micro-Chek 11:
Carrier: Epoxidized soybean oil
Density (lb/gal): 8.30
EPA #: 1486-19

Micro-Chek 11 DIDP:
Carrier: Diisodecyl phthalate
Density (lb/gal): 8.10
EPA #: 1486-19

Micro-Chek 11 S-711:
Carrier: Mixed dialkyl phthalates
Density (lb/gal): 8.20
EPA #: 1486-19

Micro-Chek 11 P:
Carrier: Organic matrix
Density (lb/gal): 9.20
EPA #: 1486-19

Micro-Chek is also available in specialty carriers for specific applications.

Microban Products Co.: Microban Antibacterial:

Microban antibacterial protection utilizes a broad spectrum antimicrobial agent effective against a wide range of gram-positive and gram-negative bacteria as well as yeast and fungi. The Microban ingredient is registered with the EPA and end-use registered for the protection of air filters and air filtration media.

Microban Additive "B" is a broad-spectrum antimicrobial chemical that has been proven effective against a variety of bacteria and fungi. The active ingredient in the Microban formulation is Triclosan, a phenolic ether compound with excellent thermal stability and chemical inertness. Coupled with its low toxicity to higher cells and organisms, it is an ideal candidate for a variety of systems requiring antimicrobial protection. The Microban additive inhibits the growth of microorganisms by utilizing an electochemical mode of action to penetrate and disrupt their cytoplasmic membranes. Upon penetration, leakage of essential metabolites from the cell occurs, resulting in a disruption of key cellular functions. These cells are then unable to reproduce and therefore eliminating the growth of the organism. Microban only affects the cell membranes of microorganisms; our mammalian cells, as well as those of all other red-blooded animals are unaffected. The four factors mode of action, low toxicity level in its pure state, low actual usage levels and binding to the product-are what makes Microban antimicrobial protection safe for use in our everyday environments.

The air filtration market holds immense opportunities for creative antimicrobial applications. Airborne contaminants, that may include potentially infectious fungal and bacteria organisms, are abundant in the environment and can lead to a variety of respiratory illnesses. These include organisms such as Aspergillus, Staphylococci and various Enterobacteriaceae. Often, air filtration media become an ideal host for the growth and propagation of fungi and bacteria. This becomes serious when conditions to sustain such organisms become optimal, such as the presence of humidity and warm temperatures. Under these conditions, microorganisms will begin to colonize the filtration substrate resulting in unpleasant odor production. Excessive colonization of these organisms will also result in the premature clogging of the air pathways of the filtration medium. It is sometimes difficult and cost inefficient to have to frequently remove air filtration media for cleaning or replacement in systems that are not easily accessible to maintenance. Therefore, there is good justification for having the added degree of protection afforded by antimicrobial elements.

There are two main ways whereby Microban can be incorporated into a variety of filtration media substrates. It may be incorporated as part of a durable topical surface treatment and it also may be incorporated directly into the substrate of interest.

Morton International, Inc.: The Right Antimicrobial: The Active Substance:

OBPA: 10,10'-Oxybisphenoxarsine:
*Recommended quantity in polymers: 200 ppm-500 ppm
*Thermal stability: approx. 300C
*Solubility in water: approx. 6 ppm
*Suitable for plast. PVC, PU, TPU, TPE, polymeric alloys, silicones, nylon....

DCOIT: Dicloro-octyl-isothiazoline:
*Recommended quantity in polymers: 800 ppm-1200 ppm
*Thermal stability: approx. 220C
*Solubility in water: approx 6 ppm
*Suitable for plast. PVC, PU, TPU, TPE, polymeric alloys, silicones....

OIT: Octyl-isothiazolinone:
*Recommended quantity in polymers: 800 ppm-1200 ppm
*Thermal stability: approx. 220C
*Solubility in water: approx. 500 ppm
*Suitable for plast. PVC, PU, TPU, TPE, polymeric alloys, silicones....

TCPP: Trichlorophenoxyphenol:
*Recommended quantity in polymers: 800 ppm - 1200 ppm
*Thermal stability: approx. 230C
*Solubility in water: approx. 10 ppm
*Suitable for plast. PVC, PU, TPU, TPE, polymeric alloys, polyolefins, silicones....

Sanqi America Inc.: APICIDER-A Series Antimicrobial:

The primary killing agent of Apacider-A is stabilized silver. The stability is achieved by bonding silver chemo-physically with calcium phosphate which is the major constituent of teeth and bone.
Apacider-A is extremely stable to light and heat, and is processed to prevent release of silver. Thus Apaciders can over come the disadvantages of other products by maximizing anti-microbial benefits while minimizing environmental impact. Apaciders withstand very high temperatures making them suitable for use with plastics and ceramics.
Apaciders are very fine white powders and remain white even after incorporation into most commonly used resins.
Apaciders demonstrate no discoloration or other time dependent changes and therefore can be made to form low maintenance surf-aces for long term use.

Special Characteristics:
  *An effective additive with broad spectrum antimicrobial
  benefits.
  Minimal-toxicity to humans or animals.
  Extremely low dissolution of silver.
  *Stable white powder
  Maintains excellent degree of whiteness after addition to
  plastics.
  *Exceptional stability to light and heat.
  *Easy admixture in most applications.

Specifications for Apacider AW:
  Appearance: Fine White Powder
  Particle Size: 1-2um (micron)
  Surface Area: 2-3 m2/g
  Heat Stability: 1,200C (2,192F)

Available Forms:
  Powders: Master Batch Pellets:
  Apacider-A35   Polypropylene
  Apacider-A25   Polyethylene
  Apacider-AW   Nylon
  Apacider-NB   Polyester
                              ABS
                              Polycarbonate
                              PVC
                              Polyol (for Polyurethane)

Sanqi America, Inc.: APACIDER-A Series Antimicrobial (Continued):

APACIDER-AW:
   Apacider-AW is an antimicrobial powder for commercial and industrial use designed to be incorporated into various materials during varying stages of the manufacturing process to preserve the integrity of the manufactured materials. A preservative for use in the manufacture of plastic films, paint (interior), paper coating, adhesive, synthetic fibers and pigments.
Active Ingredient: Silver as elemental: 2.03%
Inert Ingredients: Zinc Calcium Phosphate Silica: 97.97%
EPA Reg. No. 68317-2

APACIDER-AK:
   Apacider-AK is an antimicrobial powder for commercial and industrial use designed to be incorporated into various materials during varying stages of the manufacturing process to preserve the integrity of the manufactured materials. A preservative for use in the manufacture of plastic films, paint (interior), paper coating, adhesive, synthetic fibers and pigments.
Active Ingredient: Silver as elemental: 1.50%
Inert Ingredients: Calcium Phosphate: 98.50%
EPA Reg. No. 68317-1

Powder Specifications:
Apacider-AW:
   Particle Size (average, um): 1.5-2.5
   Particle Size (maximum, um): Under 10
   Silver Content (%): 1.8-2.5
   Zinc Content (%): 1.8-2.5
   Silver Elution (ppm): Under 0.05
   Zinc Elution (ppm): Under 0.05
   Moisture (%): Under 0.05
   Volume Density (g/cc): 0.1-0.5
   Tap Density (g/cc): 0.2-1.0
   Whiteness (spectral): Over 90
   Whiteness (KETT Method): Over 95
   Specific Gravity: 3.2
   Temperature Stability: 1,200C
   General Use: General Type

Apacider-AK:
   Particle Size (average, um): Under 0.4
   Particle Size (maximum, um): Under 3.0
   Silver Content (%): 1.0-1.5
   Silver Elution (ppm): Under 0.05
   Moisture (%): Under 1.0
   Volume Density (g/cc): 0.2-0.5
   Tap Density (g/cc): 0.4-1.0
   Whiteness (spectral): Over 90
   Whiteness (KETT Method): Over 95
   Specific Gravity: 3.2
   Temperature Stability: 1,200C
   General Use: General Type (Especially fibers)

Section VI
Bonding, Blowing and Foaming Agents

Endex Polymer Additives Inc.: ENDEX Chemical Foaming Agents and Process Aids: EXTRUSION GUIDE:

Multi-Process Application:
Endex chemical foaming agents are intended for use in a wide variety of processes, including:
* Injection molding for cycle time reduction, sink marks, warp reduction, shrink control.
* Structural Foam Molding (SFM).
* Gas Counter-Pressure foam molding.
* Low Pressure Nitrogen foam molding for nucleation and extra weight reduction.
* Gas Assist molding for extra weight reduction and sink elimination.
* Extrusion of Sheet, Film, and Profiles.

Endex Chemical Foaming Agents:
Multi-Polymer Compatability
Endex ABC 2750 & ABC 2650A are compatible with almost all polymers.

Product Information:
Endex Endothermic Chemical Foaming Agent:
ABC 2750
ABC 2650A
Physical Form: Thermoplastic polymer pellets
Active Ingredients: Acid and base components
Process temperatures: 148-315C, 300-600F
Carbon dioxide evolution: 50-55 ml/gram
Characteristics: Endothermic decomposition
Physiology: All ingredients are Generally Recognized As Safe
(GRAS) by the FDA

Endex is a 50% active pelletized concentrate in a polymer carrier. It can be used in a wide range of commodity and engineering plastics, combining quality and efficiency with benefits of versatility and cost effectiveness.

Extrusion Processes:
Expanded sheet
Co-extrusion
Blow-molding
Wire coating
Pipe and profile extrusion

Endex Polymer Additives Inc.: ENDEX Chemical Foaming Agents and Process Aids: MOLDING GUIDE:

Multi-Polymer Compatibility:
Endex is a high efficiency mini-pelletized concentrate, in a thermoplastic resin carrier. It is designed for use in both commodity and engineering thermoplastics, and may be dried with the polymer if necessary. It produces a very fine, closed cell structure, and smooth surfaces. It combines its endothermic properties, quality and efficiency, with versatility and cost effectiveness.

Endex ABC 2750 & ABC 2650A:
Endothermic Chemical Foaming Agent: Product Data:
Physical Form: Thermoplastic polymer pellets
Process temperatures: 148-315C, 300-600F
Carbon dioxide evolution, typical: 50-55 ml/gram
Characteristics: Endothermic decomposition
Physiology: All ingredients are Generally Recognized As Safe
      (GRAS) by the FDA

Multi-Process Application:
Endex endothermic chemical foaming agents are intended for use in a wide variety of processes, including:
   *Injection molding for cycle time reduction, sink marks,
   warp reduction, shrink control.
   *Structural Foam Molding (SFM).
   *Gas Counter-Pressure foam molding.
   *Low Pressure Nitrogen SFM for nucleation, void reduction,
   and extra weight reduction.
   *Gas Assist/Gas Injection molding for extra weight reduction
   and sink elimination.
   *Moisture sensitive thermoplastic polymers which need to be
   dried before processing.

Endex PBS 2625 For Moisture Sensitive Polymers:
Endex PBS 2625 is a very high efficiency, mini-pelletized concentrate, in a thermoplastic resin carrier, designed for use in very moisture sensitive engineering thermoplastics, for example, Polycarbonate and PET, and may be dried with the polymer. It produces a very fine, closed cell structure, and smooth surfaces. It combines its endothermic properties, quality and efficiency, with versatility and cost effectiveness.
Product Data:
Physical Form: Thermoplastic polymer pellets
Active Ingredients: Acid and base components
Process temperatures: 171-315C, 300-600F
Carbon dioxide evolution, typical: 25 ml/gram
Characteristics: Endothermic decomposition
Physiology: All ingredients are Generally Recognized As
      Safe (GRAS) by the FDA

Rit-Chem Co., Inc.: BLO-FOAM Chemical Blowing Agents:

Blo-Foam ADC Series:
  Type: Azodicarbonamide
  Appearance: Orange/Yellow fine powder
  Decomposition Temperature Range: 200-205C
  Gas/Volume (ml/g) @ STP: 215-225
  Average Particle Size (Fisher Subsiever):
    ADC 1200: 12 microns
    ADC 800: 8 microns
    ADC 700: 7 microns
    ADC 550: 5.5 microns
    ADC 450: 4.5 microns
    ADC 300: 3 microns
    ADC 150: 1.5 microns

Blo-Foam ADC-F Series:
  Type: Azodicarbonamide: free-flowing grade to reduce agglom-
      eration and lumping problems
  Appearance: Orange/Yellow fine powder
  Decomposition Temperature Range: 200-205C
  Gas/Volume (ml/g) @ STP: 215-225
  Average Particle Size (Fisher Subsiever):
    ADC 1200 FF: 12 microns
    ADC 800 FF: 8 microns
    ADC 700 FF: 7 microns
    ADC 550 FF: 5.5 microns
    ADC 450 FF: 4.5 microns
    ADC 300 FF: 3 microns
    ADC 150 FF: 1.5 microns

Blo-Foam ADC-FFNP Series:
  Type: Modified Azodicarbonamide for use in injection molding
      and extrusion
  Appearance: Fine yellow powder
  Decomposition Temperature Range: 195-201C
  Gas Volume (ml/g) @ STP: 230-250
  Average Particle Size (Fisher Subsiever):
    ADC 1200 FFNP: 12 microns
    ADC 800 FFNP: 8 microns
    ADC 550 FFNP: 5.5 microns
    ADC 300 FFNP: 3 microns
    ADC 150 FFNP: 1.5 microns
Blo-Foam ADC-WS Series:
  Type: Coated Azodicarbonamide for ease of dispersion in vinyl
      plastisol
  Appearance: Orange/yellow powder
  Decomposition Temperature Range: 200-205C
  Gas Volume (ml/g) @ STP: 215-225
  Average Particle Size (Fisher Subsiever):
    ADC 500 WS: 5.5 microns
    ADC 450 WS: 4.5 microns
    ADC 300 WS: 3 microns
    ADC 150 WS: 1.5 microns
    ADC 300 MC: 3 microns

Rit-Chem Co., Inc.: BLO-FOAM Chemical Blowing Agents (Continued):

Blo-Foam ADC 450 LT:
    Type: Modified Azodicarbonamide formulated for PVC foam core
        pipe and rigid PVC profiles in extrusion. Also used in
        low temperature plastisol calendering process.
    Appearance: Fine yellow powder
    Decomposition Temperature Range: 180-190C
    Gas Volume (ml/g) @ STP: 200-210
    Average Particle Size (Fisher Subsiever): 4.5 microns

Blo-Foam OBSH & Blo-Foam BBSH:
    Type: P,P',oxybisbenzene sulfonyl hydrazide. For use in
        injection molding of PP, ABS, HDPE, HIPS, and extru-
        sions of rigic PVC, PP, HIPS, ABS and HDPE.
    Appearance: Fine white powder
    Decomposition Temperature Range: 158-164C
    Gas Volume (ml/g) @ STP: 122-130

Blo-Foam SH:
    Type: p-toluene sulfonyl hydrazide. For use in expansion of
        open cell natural rubber, closed cell SBR soling,
        molded closed cell neoprene, epoxy expanding liquid
        polysulfide rubbers and polyesters.
    Appearance: Fine white powder
    Decomposition Temperature Range: 148-154C
    Gas Volume (ml/g) @ STP: 113-118

Blo-Foam RA:
    Type: p-toluene sulfonyl semicarbazide. For injection molding
        of PP, ABS, HDPE, HIPS and extrusion of rigid PVC, PP,
        HIPS, ABS, and HDPE.
    Appearance: White fine powder
    Decomposition Temperature Range: 229-235C
    Gas Volume (ml/g) @ STP: 140-150

Blo-Foam 5PT:
    Type: 5-phenyl tetrazole. For high temperature expansion of
        engineering plastics.
    Appearance: Fine white (needle-like) powder
    Decomposition Temperature Range: 234-245C
    Gas Volume (ml/g) @ STP: 190
    Melting Point: 210-220C

Blo-Foam KL Series:
    Type: Specially modified blowing agent system based on
        Azodicarbonamide. For use in low temperature thermo-
        plastic foam applications.
    Appearance: Fine yellow powder
    Decomposition Temperature Range: KL-9: 147-155C
                                      KL-10: 127-133C
                                      KL-11: 110-120C
    Gas Volume (ml/g) @ STP: KL-9: 135-145
                             KL-10: 135-145
                             KL-11: 140-150

Rit-Chem Co., Inc.: BLO-FOAM Chemical Blowing Agents (Continued):

Blo-Foam ACMP Series:
  Type: Azodicarbonamide. For the extrusion foam process of
    crosslinked polyolefins.
  Appearance: Fine yellow powder
  Decomposition Range: ACMP-F: 196-201C
                       AZO-P: 196-201C
                       AZO-F: 193-197C
                       ACMP: 193-197C
                       ACLP: 188-196C
  Gas Volume (ml/g) @ STP: ACMP-F: 270-290
                           AZO-P: 270-290
                           AZO-F: 270-290
                           ACMP: 270-290
                           ACLP: 230-250
  Average Particle Size (Coulter Counter):
    ACMP-F: 9-11.5 microns
    AZO-P: 12-14 microns
    AZO-F: 6.4-7.5 microns
    ACMP: 13-15 microns
    ACLP: 13-15 microns

Kycerol Series:
  Type: Modified sodium bicarbonate. Endothermic blowing agent
    for use in injection molding and extrusion of common
    thermoplastics.
  Appearance: White fine powder
  Melt Point: Kycerol 91: 150-160C
              Kycerol 92: 180-190C
  Gas Volume (ml/g) @ STP: Kycerol 91: 115-125
                           Kyecrol 92: 114-124
  Particle Size (Fisher Subsiever): Kycerol 91: <1 micron
                                    Kycerol 92: 3-6 microns

Section VII
Dispersants

Chemax, Inc.: MAXSPERSER Pigment Dispersants:

8913/PWDR:
    Use Level % by Weight: PE: 0.50-3.0//PP: 3.0-4.0
                           PS: 0.20-1.0//SBR: 1.0-2.0
    Form: Powder
    FDA: Yes

8953:
    Use Level % by Weight: Acrylic: 5.0-9.0
    Form: Liquid
    FDA: No

9500:
    Use Level % by Weight: PE: 0.20-0.50//PP: 0.40-0.80
    Form: Liquid
    FDA: Yes

9550:
    Use Level % by Weight: PE: 0.20-0.50//PP: 0.40-0.80
    Form: Liquid
    FDA: No

9600:
    Use Level % by Weight: PE: 0.30-0.80//PP: 0.40-0.90
    Form: Liquid
    FDA: No

9700:
    Use Level % by Weight: PE: 0.30-0.80//PP: 0.40-0.90
    Form: Liquid
    FDA: Yes

9800:
    Use Level % by Weight: PE: 0.20-0.80//PP: 0.20-0.80
    Form: Liquid
    FDA: Yes

9800/LM:
    Use Level % by Weight: PE: 0.20-0.80//PP: 0.20-0.80
    Form: Liquid
    FDA: Yes

Tego Chemie Service GmbH: TEGO Dispers Dispersing Additives:

Wetting and dispersing additives prevent flooding, floating and hard-settling of organic and inorganic pigments in water-borne and solvent-based coatings. For the manufacture of pigment pastes, Tego recommends Tego Dispers 735W/740W/745W.

Dispers 610:
  System: S
  Dosage: 0.5-2.5%
  Features: Co-flocculates inorganic pigments
Dispers 610S:
  System: S
  Dosage: 0.5-2.5%
  Features: Additional anti-floating/flooding
Dispers 630:
  System: S
  Dosage: 0.5-2.0%
  Features: Anti-sagging
Dispers 700:
  System: S
  Dosage: 0.4-4.0%
  Features: Deflocculates inorganic pigments
Dispers 710:
  System: S
  Dosage: 6.0-90.0*%
  Features: Deflocculates organic/inorganic pigments

Dispers 715W:
  System: W
  Dosage: 0.5-3.0%
  Features: Helps wetting anorganic pigments and fillers,
           strong reduction of grinding viscosity
Dispers 735W:
  System: W
  Dosage: 4.0-65.0*%
  Features: Deflocculates inorganic pigments
Dispers 740W:
  System: W
  Dosage: 2.0-70.0*%
  Features: Very good color strength development, very good
           wetting of organic and inorganic pigments, free
           of nonyl-phenol ethoxylates, FDA, solvent-free,
           cost effective
Dispers 745W:
  System: W
  Dosage: 10.0-90.0*%
  Features: Butyl glycol-free, low solvent content, for pigment
           pastes, deflocculates inorganic/organic pigments,
           very good color strength development

*high dosages are for the production of pigment pastes

Section VIII
Fillers and Extenders

Agrashell, Inc.: Vegetable Shell Products: Dual Screen Aggregates:

AD-3:
Screen Analysis:

U.S. Std. Screen No.	% Retained
12	5 max.
20	85 min.
30	99 min.

Moisture Content: 3-10%
pH Value in Water*: 4-6
Specific Gravity (typical)*: 1.3
Hardness (MOHS Scale)*: 3-4
Bulk Density-lbs./cu.ft.*: 44+-2
* Characteristics, not properties Agrashell can control

AD-4:
Screen Analysis:

U.S. Std. Screen No.	% Retained
12	1 max.
14	5 max.
30	90 min.
Pan	10 max.

Moisture Content: 3-10%

AD-6:
Screen Analysis:

U.S. Std. Screen No.	% Retained
18	5 max.
40	85 min.
60	99 min.

Moisture Content: 3-10%

AD-9:
Screen Analysis:

U.S. Std. Screen No.	% Retained
40	4 max.
100	90 min.
200	99 min.
Pan	1 max.

Moisture Content: 3-10%

AD-10.5:
Screen Analysis:

U.S. Std. Screen No.	% Retained
40	1 max.
50	2 max.
100	50 min.
200	90 min.

Moisture Content: 3-10%

Agrashell, Inc.: Industrial Flour: Made from Vegetable Shell Raw Material:

Agrashell's Industrial Flours are low specific gravity ligno-cellulose products manufactured from nut shells. The flours are light in color, non-toxic, free flowing, low in resin absorption, and easily dispersed.

Carefully manufactured to eliminate contamination, Agrashell's Industrial Flours are uniform and present no silicon hazard. Being manufactured from a "renewable resource", supply has never been a problem.

Typical Applications:

Industrial Flour is an ideal filler/extender for many synthetic resins. Some other uses are also listed below:

Rubber
Adhesives
Cosmetics
Texture Paints
Cast Polyesters
Chemical Carriers
Anti-Blocking Agents
Phenolic Molding Compounds
Cushioning Agent in Foundry Sands
Burn Out Medium in Grinding Wheels and Refractories

Properties:

Industrial Flour is available in two standard grades:
    WF-5: -200 mesh
    WF-7: -325 mesh

A minimum of ninety percent of each grade will pass through the designated U.S. Standard Sieve number. Both have the following typical properties:

Specific Gravity: 1.35
Dry Packing Density (Lbs. per cu. ft.): 40
pH Value at 25C (in Water): 5
Free Moisture (80C for 15 hrs.): 2-8%
Flash Point (Closed Cup): 380F
Oil Absorption: 55-60%
Ash Content: 0.5%

American Wood Fibers: Standard Softwood Grades:

2020:
    10 Mesh (2000 Microns): Trace
    20 Mesh (850 Microns): 0-10
    40 Mesh (425 Microns): 10-60
    60 Mesh (250 Microns): 35-90
    Balance Pan: Max 20% (thru 60 Mesh)
    Moisture Content (%): Max 8%
    Typical Bulk Density (lbs/cu.ft.): 17
    Typical Acidity (pH): 4.7
    Typical Specific Gravity: 0.4
    Typical Ash Content (%): 0.5

4020:
    20 Mesh (850 Microns): Trace
    40 Mesh (425 Microns): 0-5
    60 Mesh (250 Microns): 45-80
    80 Mesh (180 Microns): 15-45
    Balance Pan: Max 25% (thru 80 Mesh)
    Moisture Content (%): Max 8%
    Typical Bulk Density (lbs/cu.ft.): 13
    Typical Acidity (pH): 4.7
    Typical Specific Gravity: 0.4
    Typical Ash Content (%): 0.5

6020:
    40 Mesh (425 Microns): Trace
    60 Mesh (250 Microns): 0-5
    80 Mesh (180 Microns): 0-55
   100 Mesh (150 Microns): 15-40
    Balance Pan: Max 85% (Thru 100 Mesh)
    Moisture Content (%): Max 8%
    Typical Bulk Density (lbs/cu.ft.): 8
    Typical Acidity (pH): 4.7
    Typical Specific Gravity: 0.4
    Typical Ash Content (%): 0.5

8020:
    60 Mesh (250 Microns): Trace
    80 Mesh (180 Microns): 0-5
   100 Mesh (150 Microns): 0-35
   120 Mesh (125 Microns): 0-40
    Balance Pan: Min 40% (Thru 120 Mesh)
    Moisture Content (%): Max 8%
    Typical Bulk Density (lbs/cu.ft.): 8
    Typical Acidity (pH): 4.7
    Typical Specific Gravity: 0.4
    Typical Ash Content (%): 0.5

    Contaminant-Free Wood Flour:
    All products are shipped free from bark, dirt, metal and other
    foreign contamination.

American Wood Fibers: Standard Softwood Grades (Continued):

10020:
    80 Mesh (180 Microns): Trace
    100 Mesh (150 Microns): 0-5
    120 Mesh (125 Microns): Max 10%
    140 Mesh (106 Microns): Max 20%
    Balance Pan: Min 65% (Thru 140 Mesh)
    Moisture Content (%): Max 8%
    Typical Bulk Density (lbs/cu.ft.): 7
    Typical Acidity (pH): 4.7
    Typical Specific Gravity: 0.4
    Typical Ash Content (%): 0.5

12020:
    100 Mesh (150 Microns): Trace
    120 Mesh (125 Microns): 0-5
    140 Mesh (106 Microns): Max 10%
    200 Mesh (75 Microns): Max 40%
    Balance Pan: Min 55% (Thru 200 Mesh)
    Moisture Content (%): Max 8%
    Typical Bulk Density (lbs/cu.ft.): 7
    Typical Acidity (pH): 4.7
    Typical Specific Gravity: 0.4
    Typical Ash Content (%): 0.5

14020:
    120 Mesh (125 Microns): Trace
    140 Mesh (106 Microns): 0-5
    200 Mesh (75 Microns): Max 20%
    Balance Pan: Min 65% (Thru 200 Mesh)
    Moisture Content (%): Max 8%
    Typical Bulk Density (lbs/cu.ft.): 8
    Typical Acidity (pH): 4.7
    Typical Specific Gravity: 0.4
    Typical Ash Content (%): 0.5

    Contaminant-Free Wood Flour:
    All products are shipped free from bark, dirt, metal and other
    foreign contamination.

American Wood Fibers: Standard Hardwood Grades:

2010:
    10 Mesh (2000 Microns): Trace
    20 Mesh (850 Microns): 0-15
    40 Mesh (425 Microns): 45-90
    60 Mesh (250 Microns): 5-55
    Balance Pan: 0-15 (Thru 60 Mesh)
    Moisture Content (%): Max 6%
    Typical Bulk Density (lbs/cu.ft.): 21
    Typical Acidity (pH): 5.0
    Typical Specific Gravity: 0.54
    Typical Ash Content (%): 0.7

4010:
    20 Mesh (850 Microns): Trace
    40 Mesh (425 Microns): 0-5
    60 Mesh (250 Microns): 25-85
    80 Mesh (180 Microns): 10-65
    Balance Pan: 0-25 (Thru 80 Mesh)
    Moisture Content (%): Max 6%
    Typical Bulk Density (lbs/cu.ft.): 15
    Typical Acidity (pH): 5.0
    Typical Specific Gravity: 0.54
    Typical Ash Content (%): 0.7

6010:
    40 Mesh (425 Microns): Trace
    60 Mesh (250 Microns): 0-15
    80 Mesh (180 Microns): 0-45
   100 Mesh (150 Microns): 5-40
    Balance Pan: 15-85 (Thru 100 Mesh)
    Moisture Content (%): Max 6%
    Typical Bulk Density (lbs/cu.ft.): 14
    Typical Acidity (pH): 5.0
    Typical Specific Gravity: 0.54
    Typical Ash Content (%): 0.7

8010:
    60 Mesh (250 Microns): Trace
    80 Mesh (180 Microns): 0-15
   100 Mesh (150 Microns): 0-45
   120 Mesh (125 Microns): 0-40
    Balance Pan: Min 30% (Thru 120 Mesh)
    Moisture Content (%): Max 6%
    Typical Bulk Density (lbs/cu.ft.): 13
    Typical Acidity (pH): 5.0
    Typical Specific Gravity: 0.54
    Typical Ash Content (%): 0.7

Contaminant-Free Wood Flour
All products shipped are free from bark, dirt, metal and other
foreign contamination.

American Wood Fibers: Standard Hardwood Grades (Continued):

10010:
  80 Mesh (180 Microns): Trace
  100 Mesh (150 Microns): 0-5
  120 Mesh (125 Microns): 0-20
  140 Mesh (106 Microns): 0-40
  Balance Pan: Min 45% (Thru 140 Mesh)
  Moisture Content (%): Max 7%
  Typical Bulk Density (lbs/cu.ft.): 12
  Typical Acidity (pH): 5.0
  Typical Specific Gravity: 0.54
  Typical Ash Content (%): 0.7

12010:
  100 Mesh (150 Microns): Trace
  120 Mesh (125 Microns): 0-5
  140 Mesh (106 Microns): 0-20
  200 Mesh (75 Microns): 0-45
  Balance Pan: Min 35% (Thru 200 Mesh)
  Moisture Content (%): Max 7%
  Typical Bulk Density (lbs/cu.ft.): 12
  Typical Acidity (pH): 5.0
  Typical Specific Gravity: 0.54
  Typical Ash Content (%): 0.7

14010:
  120 Mesh (125 Microns): Trace
  140 Mesh (106 Microns): 0-10
  200 Mesh (75 Microns): 0-55
  Balance Pan: Min 40% (Through 200 Mesh)
  Moisture Content (%): Max 7%
  Typical Bulk Density (lbs/cu.ft.): 11
  Typical Acidity (pH): 5.0
  Typical Specific Gravity: 0.54
  Typical Ash Content (%): 0.7

Contaminant-Free Wood Flour
All products shipped are free from bark, dirt, metal and other foreign contamination.

American Wood Fibers: Wood Flour:

Maple (Hardwood):
   Plant Location: Schofield, WI
   Grade: 8 Standard (20-200 Mesh)
   Typical Aspect Ratio (l/w): 3/1
   Typical Moisture Content (%): 3-7%
   Typical Moisture Absorption ("x" times wgt. wt): 4.0
   Typical Cell Content: Tannic Acid
   Typical pH: 5.0
   Typical pH in Solution: 3.7
   Typical Bulk Density (lbs/cuft): 10-18
   Typical Specific Gravity: 0.54
   Typical Ash Content: 0.2
   Typical Color: Red/Brown

Ponderosa Pine (Softwood):
   Plant Location: Schofield, WI
   Grade: 7 Standard (20-140 Mesh)
   Typical Aspect Ratio (l/w): 4/1
   Typical Moisture Content (%): 8% Maximum
   Typical Moisture Absorption ("x" times wgt. wt): 5.5
   Typical Cell Content: Resin
   Typical pH: 4.7
   Typical pH in solution: 3.4
   Typical Bulk Density (lbs/cuft): 7-17
   Typical Specific Gravity: 0.4
   Typical Ash Content: 0.5
   Typical Color: Buff

Ponderosa Pine (Softwood):
   Plant Location: Pella, IA
   Grade: 5 Standard (10, 20, 40, 80, 100 Mesh)
   Typical Aspect Ratio (l/w): 4/1
   Typical Moisture Content (%): 8% Maximum
   Typical Moisture Absorption ("x" times wgt. wt): 5.5
   Typical Cell Content: Resin
   Typical pH: 4.7
   Typical pH in Solution: 3.4
   Typical Bulk Density (lbs/cuft): 7-17
   Typical Specific Gravity: 0.4
   Typical Ash Content: 0.5
   Typical Color: Buff

Ponderosa Pine (Softwood):
   Plant Location: Marysville, CA
   Grade: 2 Standard (10, 20 Mesh) (40 Mesh-Conditional)
   Typical Aspect Ratio (l/w): 4/1
   Typical Moisture Content (%): 8% Maximum
   Typical Moisture Absorption ("x" times wgt. wt): 5.5
   Typical Cell Content: Resin
   Typical pH: 4.7
   Typical pH in Solution: 3.4
   Typical Bulk Density (lbs/cuft): 13-17
   Typical Specific Gravity: 0.4
   Typical Ash Content: 0.5
   Typical Color: Buff

American Wood Fibers: Wood Flour (Continued):

Oak (Hardwood):
  Plant Location: Struthers, OH
  Grade: 3 Standard (10, 20, & 40 Mesh)
  Typical Aspect Ratio (l/w): 2.5-3/1
  Typical Moisture Content (%): 3-5%
  Typical Moisture Absorption ("x" times wgt. wt): 4.3
  Typical Cell Content: >Tannic Acid
  Typical pH: 3.5
  Typical pH in Solution: 3.3
  Typical Bulk Density (lbs/cuft): 12-18
  Typical Specific Gravity: 0.63
  Typical Ash Content: 0.4
  Typical Color: Lgt. Brown

Lobolly Southern Pine (Softwood):
  Plant Location: Jessup, MD
  Grade: 3 Standard (10, 20, & 40 Mesh)
  Typical Aspect Ratio (l/w): 4/1
  Typical Moisture Content (%): 5-10%
  Typical Moisture Absorption ("x" times wgt. wt): 4.7
  Typical Cell Content: >Resin
  Typical pH: 4.7
  Typical pH in Solution: 3.4
  Typical Bulk Density (lbs/cuft): 8-16
  Typical Specific Gravity: 0.51
  Typical Ash Content: 0.2
  Typical Color: Buff/Yellow

Spruce (Softwood):
  Plant Location: Struthers, OH
  Grade: 3 Standard (20, & 40 Mesh)
  Typical Aspect Ratio (l/w): 4/1
  Typical Moisture Content (%): 12-18% (10, 20 Mesh)
                                     4- 9% process (40 mesh)
  Typical Moisture Absorption ("x" times wgt. wt): 5.2
  Typical Cell Content: Resin
  Typical pH: 7.0
  Typical pH in Solution: 3.9
  Typical Bulk Density (lbs/cuft): 6-14
  Typical Specific Gravity: 0.36
  Typical Ash Content: 0.3
  Typical Color: Off White

Claremont Flock Corp.: Precision and Random Cut Flock:

Acrylic:
Precision and Random Cut Flock
Acrylic Fiber: Synthetic fiber spun from polymers consisting
of at least 85% by weight acrylonitrile units.
Available Product Forms:
Denier (dpf): 1.5, 2.0, 3.0
Fiber Length: Precision Cut: from 0.020" to 0.125"/Random
Luster: Dull, Bright
Shape: Round, Bean, Bone
Finish: Mechanical, Electrostatic (AC or DC)
Technical Properties:
Tensile Strength: 30,000 to 45,000 lb/sq.in.
Moisture Regain: 1.0-3.5% (70F 65%rh)
Effect of Heat: Does not melt; Sticks to metal at 215 to 255C
Acceptance of Dye: Receptive to most dyestuff; Available
solution dyed colors.
Resistance to Sunlight: Natural fibers are excellent.
Dyed fibers are dependent on the dyestuffs used.
General: Good abrasion resistance. Resistant to mildew.
Not affected by mild acids and alkalis.
Typical Applications:
Decorative Ribbon
Packaging
Floor Coverings
Paint Rollers
Thermal Blankets

Cotton:
Random Cut Flock
Cotton Flock: Short-cut cotton fiber of various lengths produced
from textile mill by-products, thread waste or virgin staple
cotton fiber.
Available Product Forms:
Denier (dpf): Variable: 1.3-2.0
Fiber Length: Random: 80% between 0.015" and 0.050"
Luster: Natural
Shape: Variable: Generally Round or Oval, often irregular
Finish: Mechanical, Wet Dispersion, Electrostatic (AC or DC)
Technical Properties:
Tensile Strength: 40,000 to 120,000 lb/sq in
Moisture Regain: 6-8% (70F 60% rh)
Effect of Heat: Yellows at 120C. Decomposes at 150C.
Acceptance of Dye: Receptive of a wide variety of dyestuffs.
Resistance to Sunlight: Prolonged exposure causes loss of
strength and yellowing.
General: Good abrasion resistance; Attacked by mildew; Good
adhesion to natural and synthetic rubber compounds;
Degrades in hot acids; Excellent resistance to alkalis.
Typical Applications:
Drapery Backing
Latex Glove Lining
Filler/Strengthening Agent for Rubber Compounds
Colorant Fibers for Paper and Plastic

Claremont Flock Corp.: Precision and Random Cut Flock (Continued):

Nylon (Polyamide Types 6 and 6,6):
Precision and Random Cut Flock
Synthetic fiber composed of a long chain of polyamide in
which less than 85% of the amide linkages are attached to two
atomic rings.
Available Product Forms:
Denier (dpf): 0.8,1.0,1.3, 1.5, 1.8, 2.0, 3.0, 6.0, 18.0, 20.0
Fiber Length: Precision Cut: from 0.0150" to 0.250"/Random
Luster: Dull, Semi-Dull, Bright
Shape: Round, Trilobal
Finish: Mechanical, Electrostatic (AC or DC)
Technical Properties:
Tensile Strength: 40,000 to 106,000 lb/sq in
Moisture Regain: 4.0-4.5% (70F 65% rh)
Effect of Heat: Receptive to a wide range of dyestuffs
Resistance to Sunlight: Prolonged exposure causes degradation
    and yellowing.
General: Excellent abrasion resistance; Resistant to mildew;
    Decomposes in strong acids; Unaffected by alkali.
Typical Applications:
Upholstery/Home Furnishings
Apparel
Decorative Applications (Decals, Greeting Cards)
Reinforcing Filler in Rubber Compounds
Paint Rollers
Thermal Blankets

Polyester:
Precision Cut Flock
Synthetic fiber spun from long chain polymers composed of at
least 85% by weight of an ester such as terephthalate.
Available Product Forms:
Denier (dpf): 1.0, 1.5, 3.0
Fiber Length: Precision Cut: from 0.015" to 0.250"
Luster: Dull, Semi-Dull; Bright
Shape: Round
Finish: Electrostatic (AC or DC)
Technical Properties:
Tensile Strength: 50,000 to 99,000 lb/sq in
Moisture Regain: 0.4% (70F 60% rh)
Effect of Heat: Sticks at 230C; Melts at 260C
Acceptance of Dye: Disperse and azoic dyestuffs recommended;
Available in a wide variety of solution dyed colors, including
black for automotive applications.
Resistance to Sunlight: Generally excellent, however prolonged
exposure causes some loss of strength.
General: Good abrasion resistance; Excellent resistance to
mildew; Good resistance to acids; Less resistant to alkali.
Typical Applications:
Automotive Interior Structures
Automotive Window Channel
Filler for Reinforced Plastic and Rubber

Claremont Flock Corp.: Precision and Random Cut Flock (Continued):

Viscose Rayon:
Precision and Random Cut Flock
A manufactured fiber composed of regenerated cellulose in which substitutes have replaced not more than 15% of the hydrogens of the hydroxyl groups.
Available Product Forms:
Denier (dpf): 0.9, 1.5, 3.0, 5.0, 25.0
Fiber Length: Precision Cut: from 0.020" to 0.250"/Random
Luster: Dull, Bright
Shape: Round
Finish: Mechanical, Wet Dispersion, Electrostatic (AC or DC)
Technical Properties:
Tensile Strength: 30,000 to 46,000 lb/sq in
Moisture Regain: 11% (70F 60% rh)
Effect of Heat: Decomposes at 185 to 205C; does not melt or become tacky.
Acceptance of Dye: Receptive to all cotton dyestuffs; Available in a limited number of solution dyed colors.
Resistance to Sunlight: No discoloration; Long term exposure causes loss of tensile that is worse as the TiO2 level increases.
General: Good abrasion resistance; Attacked by mildew; Acids and alkalis may weaken or swell fibers.
Typical Applications:
Apparel
Packaging
Optical Polishing
Wall Coverings
Cosmetics
Colorant Fibers for Paper and Plastic

Dequssa AG: Aerosil R 972 as Filler in Fluoroelastomers:

Physico-chemical data of hydrophobic Aerosil R972:
  BET surface area: m2/g: 110+-20
  Average primary particle size: nm: 16
  Tapped density:
    Normal material: g/l: approx. 50
    Densed material: g/l: approx. 90
  Moisture: %: <0.5
  Ignition loss: %: <2
  C content: %: 1.0
  Methanol wettability: %: 40
  pH value: 3.6-4.3
  Refractive index: 1.45
  Density: g/cm3: 2.2

Aerosil is a very light, bluish-white powder which is built up of very finely divided primary particles. These primary particles are not present in a condition isolated from each other. Aerosil consists of small aggregates which join to form agglomerates. Aerosil is produced in a number of grades with varying specific surface areas.

Hydrophobic Silicas:
  Hydrophobic silicas are aftertreated products. Aftertreatment processes can be carried out with the various Aerosil types as well as with the different precipitated silicas. Aerosil R 972 has been offered on the market since 1962. It is the oldest chemically aftertreated synthetic silica, i.e. the first hydrophobic product. In contrast to the silicas which are hydrophilic by nature, the hydrophobic variants are not wetted by water. Despite the higher density of these hydrophobic silicas in comparison with water, they float on the water surface.

Survey of the use of Aerosil R 972 in fluoroelastomers:
1. 1980: Polymer Used: Fluorel 2460
   R 972 contained in all formulations
2. 1982: Polymer Used: Aflas 150 P and others
   R 972 improves stability with respect to NACE
3. 1982: Polymer Used: Aflas types
   R 972 produces high elongation at break after aging
4. 1983: Polymer Used: Aflas types
   R 972 improves "API Extrusion resistance"
5. 1984: Polymer Used: Viton B & GLT, Aflas 150 P
   R 972 + Al2O3 produces good elasticity
6. 1984: Polymer Used: Various duPont types
   R 972 + MT black produces best extrusion strength
7. 1984: Polymer Used: Various duPont types
   R 972 produces high modulus with good elongation
8. 1985: Polymer Used: Aflas
   R 972 shows good aging characteristics

D.J. Enterprises, Inc.: SILLUM-200-Q/P Alumina Silicate:

Sillum-200 Q/P is a non-combustible, chemically stable inert
alumina silicate, which is processed to eliminate any organic
tramp elements which could modify performance.
Processing the material to a free-flowing bead standard 44-75
micron screen analysis provides for ease of blending with liquid
or powder materials.

Typical Analysis:

SiO2:	62%	S.G.:	2.35
Al2O3:	37%	pH:	6.0-7.0
Na:	0.50%	Moisture:	0.05
Fe:	0.50%	Melt Point:	2,600 F+
C:	0.15%	Bulk Dens:	50-60#/cu.ft.

Free silica (Quartz) is expected only at trace levels

Properties:
Stable-insoluble
Non-thixotropic
High filler loading capability 40-60%
Excellent suspension & dispersion qualities
Water resistant
Low moisture content
Non-hygroscopic
Reduce shrinkage
Improve hardness
Reduce exotherm temperature
Lower viscosity
Compatible with ATH

ECC International: Calcium Carbonate:

SUPERMITE*:
An ultrafine ground pigment developing superior physical properties and surface gloss in plastics, elastomers and coatings applications.
Particle Size Range (micrometers): Up to 8
Mean Particle Size (micrometers): 1.0
Oil Absorption (rub-out): 18 to 21

MICRO-WHITE 15*:
An ultrafine wet-ground product for the paint and polymer industries. Used where finer products are not required but where a low top size is needed.
Particle Size Range (micrometers): Up to 10
Mean Particle Size (micrometers): 2.0
Oil Absorption (rub-out): 16 to 18

ATOMITE/MICRO-WHITE 25*:
A highly developed easy dispersing pigment that meets the most exacting requirements demanded by the paint, rubber, plastics, floor covering, and many other industries.
Particle Size Range (micrometers): Up to 15
Mean Particle Size (micrometers): 3.0
Oil Absorption (rub-out): 14 to 16

SNOWFLAKE WHITE:
A general purpose easy dispersing pigment that has found wide use in practically all types of protective coatings, rubber, plastics, caulks, glazing compounds, mastics and other end products.
Particle Size Range (micrometers): Up to 25
Mean Particle Size (micrometers): 5.5
Oil Absorption (rub-out): 10 to 12

SNOWFLAKE P.E.:
A medium ground pigment with the theoretically correct distribution of particles to ensure maximum loading to highly filled systems: e.g., polyester resins, SMC, BMC, TMC, XMC.
Particle Size Range (micrometers): Up to 44
Mean Particle Size (micrometers): 5.5
Oil Absorption (rub-out): 8-10

DRIKALITE:
A competitive grade of extender pigment for use in places where a maximum number of fine particles per unit cost is imperative.
Particle Size Range (micrometers): Up to 44
Mean Particle Size (micrometers): 7.0
Oil Absorption (rub-out): 8-10

* Registered with NSF

ECC International, Inc.: Calcium Carbonate (Continued):

DURAMITE:
A pigment with unique particle size distribution. Its easy dispersion characteristics have demonstrated its unusual merit as a mix-in type of pigment for a wide variety of applications such as carpet backing, roof compounds, spackles, and coatings.
Particle Size Range (micrometers): Up to 44
Mean Particle Size (micrometers): 11.0
Oil Absorption (rub-out): 7-9

No. 1 White:
A general purpose pigment for many uses where fineness is of secondary importance.
Particle Size Range (micrometers): Up to 50
Mean Particle Size (micrometers): 13.0
Oil Absorption (rub-out): 7-9

MICRO-WHITE 100:
A white general purpose filler for paints, rubber goods, putties and joint compounds.
Particle Size Range (micrometers): Up to 100
Mean Particle Size (micrometers): 17.0

CP FILLER:
A competitively priced filler for use in construction products where color is of secondary importance.
Particle Size Range (micrometers): Up to 100
Mean Particle Size (micrometers): 17.0

MARBLE DUST:
A coarse ground natural calcium carbonate for use in putties, glazes and mild abrasive compounds. It is frequently blended with Atomite to achieve a wider distribution of particles which yields effective particle packing properties.
Particle Size Range (micrometers): Up to 150
Mean Particle Size (micrometers): 21.0

CC-103:
A series of coarse ground calcium carbonate fillers used in polyolefins, carpet backing, caulks, sealants, and putties, and as mild abrasives in cleaners. These products offer a formulator a choice of fillers having low oil absorption, low resin demand, and average dry brightness.
Particle Size Range (micrometers): Up to 800
Mean Particle Size (micrometers): 80.0

MARBLEMITE:
A high brightness calcium carbonate product made from hand-picked stone designed for high loading and minimum black specks for cultured marble applications.
Particle Size Range (micrometers): Up to 800
Mean Particle Size (micrometers): 80.0

ECC International Inc.: Calcium Carbonate (Continued):

Surface Modified Products:
SUPERCOAT*:
A coated ultrafine ground pigment offering easy incorporation and dispersion in plastics applications which result in improved physical properties, improved production throughput efficiency with low abrasiveness.
Particle Size Range (micrometers): Up to 8
Mean Particle Size (micrometers): 1.0
Oil Absorption (rub-out): 15-17

OPACICOAT*:
An ultrafine ground coated pigment with a unique particle size distribution. The clean top size and lack of ultrafine should permit higher loadings in polymer systems while still maintaining properties.
Particle Size Range (micrometers): Up to 8
Mean Particle Size (micrometers): 1.1
Oil Absorption (rub-out): 15-17

MICRO-WHITE 15 SAM*:
An ultrafine ground coated pigment for applications requiring superior physical properties and economically priced filler.
Particle Size Range (micrometers): Up to 10
Mean Particle Size (micrometers): 2.0
Oil Absorption (rub-out): 13-15

KOTAMITE*:
A coated pigment designed for easy dispersion in plastic compounds, e.g., polyolefins, rigid and flexible PVC, Hydrophobic for superior wire and cable insulation compounds and improved impact properties in polypropylene.
Particle Size Range (micrometers): Up to 15
Mean Particle Size (micrometers): 3.0
Oil Absorption (rub-out): 12-14

* Registered with NSF

ECC International Inc.: Calcium Carbonate (Continued):

Slurry Products:
MICRO-WHITE 07*:
An ultrafine wet-ground product available for applications requiring high gloss and opacity in coatings and inks.
Particle Size Range (micrometers): up to 5
Mean Particle Size (micrometers): 0.7

Micro-White 10:
An ultrafine ground pigment to provide improved opacity and gloss properties in water based paints and coatings.
Particle Size Range (micrometers): Up to 8
Mean Particle Size (micrometers): 1.0

Micro-White 15:
An ultrafine slurry used for water based paints and inks.
Particle Size Range (micrometers): Up to 10
Mean Particle Size (micrometers): 2.0

Micro-White 25:
A fine ground carbonate slurry for use in coatings systems where ultrafine particle size is not required.
Particle Size Range (micrometers): Up to 15
Mean Particle Size (micrometers): 3.0

* Registered with NSF

ECC International Inc.: Calcium Carbonate (Continued):

CAMEL-CAL*:
An ultrafine ground pigment that has demonstrated improved physical properties in plastics, elastomers, and coating applications.
Particle Size Range (micrometers): Up to 5
Mean Particle Size (micrometers): 0.7
Oil Absorption (rub-out): 18 to 23

CAMEL-FINE*:
An ultrafine ground pigment for the paint and polymer industry. Used where finer products are not required but where a low top size is needed.
Particle Size Range (micrometers): Up to 10
Mean Particle Size (micrometers): 2.0
Oil Absorption (rub-out): 16 to 18

CAMEL-WITE*:
A fine ground pigment that has been a standard for the coatings, plastics, rubber and other industries.
Particle Size Range (micrometers): Up to 12
Mean Particle Size (micrometers): 3.0
Oil Absorption (rub-out): 14-18

CAMEL-TEX:
A medium fine ground, general purpose pigment that has found wide use in all types of coatings, rubber, plastics, caulks, glazing compounds and mastic applications.
Particle Size Range (micrometers): Up to 25
Mean Particle Size (micrometers): 4.0-5.0
Oil Absorption (rub-out): 12-16

CAMEL-FIL:
A medium ground pigment that has been designed specifically for the glass reinforced polyester industry; particle distribution controls viscosity.
Particle Size Range (micrometers): Up to 44
Mean Particle Size (micrometers): 5.5-6.0
Oil Absorption (rub out): 13-17

CAMEL-CARB:
A medium ground pigment for use in applications where a maximum number of fine particles per unit cost is important.
Particle Size Range (micrometers): Up to 44
Mean Particle Size (micrometers): 7.0
Oil Absorption (rub out): 10-14

* Registered with NSF

ECC International, Inc.: Calcium Carbonate (Continued);

C-55:
A general purpose extender pigment that has applications in coatings, caulks, sealants, rubber and plastics industries.
Particle Size Range (micrometers): Up to 44
Mean Particle Size (micrometers): 8.5
Oil Absorption (rub-out): 9-14

GSP-30:
A general purpose product for joint compound, putties, rubber goods and coatings applications.
Particle Size Range (micrometers): Up to 100
Mean Particle Size (micrometers): 16

GSP-40:
A general purpose pigment for uses where fineness is of secondary importance.
Particle Size Range (micrometers): Up to 50
Mean Particle Size (micrometers): 12.5
Oil Absorption (rub-out): 9-13

GSP-80:
A medium fine ground, easily dispersed pigment widely used in the coatings, caulks, adhesives and rubber industries.
Particle Size Range (micrometers): Up to 20
Mean Particle Size (micrometers): 4.0-5.0
Oil Absorption (rub-out): 13-17

GSP-95:
A fine ground, easily dispersed pigment for the coatings, caulks, adhesives, rubber and flexible PVC industries.
Particle Size Range (micrometers): Up to 15
Mean Particle Size (micrometers): 3.3
Oil Absorption (rub-out): 14-18

ECC International Inc.: Calcium Carbonate (Continued):

Surface Modified Products:
CAMEL-CAL ST*:
   A coated, ultrafine ground pigment that provides increased impact strength and reinforcement with excellent dispersion in rigid PVC applications.
   Particle Size Range (micrometers): Up to 5
   Mean Particle Size (micrometers): 0.7
   Oil Absorption (rub-out): 16-22

CAMEL-FINE ST*:
   A coated, ultrafine ground pigment that offers an excellent balance of impact strength, reinforcement, excellent dispersion, and cost in several thermoplastic applications.
   Particle Size Range (micrometers): Up to 10
   Mean Particle Size (micrometers): 2.0
   Oil Absorption (rub-out): 16-20

CAMEL-WITE ST*:
   A coated, fine ground pigment that provides cost effective reinforcement and excellent dispersion in several thermoplastic applications.
   Particle Size Range (micrometers): Up to 12
   Mean Particle Size (micrometers): 3.0
   Oil Absorption (rub-out): 16-20
* Registered with NSF

Slurry Products:
CAMEL-CAL Slurry:
   An ultrafine wet ground product available for applications requiring high gloss.
   Particle Size Range (micrometers): Up to 5
   Mean Particle Size (micrometers): 0.7

CAMEL-FINE Slurry:
   An ultrafine slurry for water based paints and inks.
   Particle Size Range (micrometers): Up to 10
   Mean Particle Size (micrometers): 2.0

CAMEL-WITE Slurry:
   A fine ground slurry for use in systems where ultrafine particle size is not required.
   Particle Size Range (micrometers): Up to 12
   Mean Particle Size (micrometers): 3.0

Fiber Sales & Development Corp.: SOLKA-FLOC Natural Filler:

Cellulose is the building block of all plant structure, a completely renewable resource.

Filler Attributes:
*Insoluble in Water and Organic Solvents
*Resistant to Dilute Acids and Alkalies
*Viscosity Control
*Cracking Inhibitor
*Shrinkage Control
*Improved Adhesive Strength
*Longer Open Time
*Increased Volume Coverage

Filler Uses:
*Reinforcing Filler
*Bulking Agent
*Thickening Aid
*Plasticizer
*Arc Intensifier

*Texturizing Aid
*Processing Aid
*Conditioning Agent
*Lubricant
*Absorbent

Filler Applications:
*Extruded and Calendered
 Goods
*Molded Goods
*Proofed Goods
*Blown Sponge
*Roll and Tank Liners
*Latex Adhesives
*Ceramics

*Latex Paints
*Texture and Adhesive Compounds
*Electrical Products
*Rubber and Plastic Products
*Floor Tiles
*Shoe Soles and Heels
*Tires
*Welding

Typical Properties:
Grade:
1016:
Average Fiber Length u: 290
Bulk Volume (cc/g): 5.5-6.5
10:
Average Fiber Length u: 120
Bulk Volume (cc/g): 5.2-6.2
20:
Average Fiber Length u: 100
Bulk Volume (cc/g): 3.0-4.0
40:
Average Fiber Length u: 65
Bulk Volume (cc/g): 2.8-3.3
100:
Average Fiber Length u: 40
Bulk Volume (cc/g): 2.2-2.6
300:
Average Fiber Length u: 22
Bulk Volume (cc/g): 2.1-2.4

Fibertec: MICROGLASS Milled Fibers:

Microglass milled fibers are E-Glass filaments hammermilled to various bulk densities.

Microglass is used as a filler/reinforcement in composites to increase mechanical properties (tensile, flexural and impact), improve dimensional stability and minimize distortion at elevated temperatures.

Product Sizing and End Use:
  3000 Series--Unsized fibers compatible with all thermoset and thermoplastic resins.
  6000 Series--Low cost fibers with higher levels of sizing compatible with all thermoset and thermoplastic resins.
  7000 Series--Cationic sized to accelerate wet out, improve dispersion, and overall processability. Compatible with all thermoset and thermoplastic resins.
  9000 Series--Silane sized fibers for improved bonding between the inorganic glass and the organic resin. Sizings are available for polyesters, epoxies, phenolics, polyurethanes, polyvinyl chloride and other resin systems.

Milled Fiber Specifications
Unsized

3082:
    Screen Size (inches): 1/32
    Average Fiber Diameter (microns): 16
    Average Fiber Length (microns): 120
    Appearance: Powdery
3080:
    Screen Size (inches): 1/32
    Average Fiber Diameter (microns): 16
    Average Fiber Length (microns): 150
    Appearance: Powdery
3032:
    Screen Size (inches): 1/32
    Average Fiber Diameter (microns): 16
    Average Fiber Length (microns): 200
    Appearance: Powdery
3016:
    Screen Size (inches): 1/16
    Average Fiber Diameter (microns): 10
    Average Fiber Length (microns): 140
    Appearance: Floccular
3004:
    Screen Size (inches): 1/4
    Average Fiber Diameter (microns): 10
    Average Fiber Length (microns): 210
    Appearance: Floccular

Fibertec: MICROGLASS Milled Fibers (Continued):

Milled Fiber Specifications (Continued)
Cationic

7280:
Screen Size (inches): 1/32
Average Fiber Diameter (microns): 16
Average Fiber Length (microns): 120
Appearance: Powdery

7232:
Screen Size (inches): 1/32
Average Fiber Diameter (microns): 16
Average Fiber Length (microns): 230
Appearance: Powdery

7216:
Screen Size (inches): 1/16
Average Fiber Diameter (microns): 10
Average Fiber Length (microns): 170
Appearance: Floccular

7204:
Screen Size (inches): 1/4
Average Fiber Diameter (microns): 10
Average Fiber Length (microns): 230
Appearance: Floccular

Silane

9110:
Screen Size (inches): 1/32
Average Fiber Diameter (microns): 16
Average Fiber Length (microns): 150
Appearance: Powdery

9132:
Screen Size (inches): 1/32
Average Fiber Diameter (microns): 16
Average Fiber Length (microns): 220
Appearance: Powdery

9007D:
Screen Size (inches): 1/16
Average Fiber Length (microns): 10
Average Fiber Length (microns): 150
Appearance: Floccular

9114:
Screen Size (inches): 1/4
Average Fiber Length (microns): 10
Average Fiber Length (microns): 160
Appearance: Floccular

Fibertec: MICROGLASS MIlled Fibers (Continued):

Milled Fiber Specifications (Continued)
Direct Mill

6632:
Screen Size (inches): 1/32
Average Fiber Diameter (microns): 16
Average Fiber Length (microns): 170
Appearance: Powdery

6608:
Screen Size (inches): 1/8
Average Fiber Diameter (microns): 16
Average Fiber Length (microns): 470
Appearance: Floccular

6616:
Screen Size (inches): 1/16
Average Fiber Diameter (microns): 16
Average Fiber Length (microns): 290
Appearance: Floccular

Fibertec: Wollastonite (Calcium Metasilicate):

Fibertec Wollastonites are acicular (fibrous or needle-like) mineral products milled to various bulk densities and particle size distributions. These products can be surface treated to enhance performance in selected resin systems.

Fibertec Wollastonites are used as functional fillers and reinforcements in composites (thermosets) and compounds (thermoplastics) to increase mechanical properties (tensile, flexural, impact), minimize distortion at elevated temperatures, improve dimensional stability, and provide low moisture absorption.

Chemical Composition-All Products:

Component:	% by Weight
CaO	46.70
SiO2	52.20
Fe2O3	0.15
Al2O3	0.25
MgO	0.20
L.O.I. (inorganic)	0.50

Functional Properties and Morphology:

520HD:
    Average Diameter (microns): 5
    Average Aspect Ratio: 20
    Loose Bulk Density (gms/cc): 0.35
    Tapped Bulk Density (gms/cc): 0.58
    Appearance: Acicular white particle

905U:
    Average Diameter (microns): 9
    Average Aspect Ratio: 5
    Loose Bulk Density (gms/cc): 0.75
    Tapped Bulk Density (gms/cc): 1.16
    Appearance: Powdery

915SH:
    Average Diameter (microns): 9
    Average Aspect Ratio: 15
    Loose Bulk Density (gms/cc): 0.52
    Tapped Bulk Density (gms/cc): 0.90
    Appearance: Acicular white particle

1515U:
    Average Diameter (microns): 15
    Average Aspect Ratio: 15
    Loose Bulk Density (gms/cc): 0.50
    Tapped Bulk Density (gms/cc): 0.90
    Appearance: Acicular white particle

Grefco Minerals, Inc.: DIATOMITE Functional Fillers:

Dicalite Diatomite functional fillers are produced from
diatomaceous earth, a versatile and valuable raw material.
Diatomite is more than just a crystal or mineral that formed
in a rock. It consists of delicately constructed silica
skeletons grown by uncounted microscopic organisms each with
its own design. Deposits of these skeletons are collections of
solid and perforated rods, disks, hemispheres, crescents and
polygons. Because of their unusual physical structure the
particles interlace and overlay in a random, three dimensional
matrix which stiffens, reinforces and improves the durability
of filled systems. This myriad of shapes also offers major
advantages in terms of low density and high absorption.
Natural Diatomite Functional Fillers:
   Color: Cream
   Moisture, %: 6.0
   Ig. Loss (dry basis) %: 5.0
   pH (10% slurry): 6.5-8.5

104:
   Brightness (G.E.): 76
   Oil Absorption (GCOA), wt%: 160
   *Effective Density g/cc: 2.4
   Hegman Grind: 5.0
   MPD (micron): 3.7
CA-3:
   Brightness (G.E.): 71
   Oil Absorption (GCOA), wt%: 170
   *Effective Density g/cc: 2.3
   Hegman Grind: 4.5
   MPD (micron): 6.6
IG-3:
   Brightness (G.E.): 70
   Oil Absorption (GCOA), wt%: 170
   *Effective Density g/cc: 2.2
   Hegman Grind: 4.5
   MPD (micron): 7.1
143:
   Brightness (G.E.): 72
   Oil Absorption (GCOA), wt%: 175
   *Effective Density g/cc: 2.2
   Hegman Grind: 4.5
   MPD (micron): 7.4
SA-3:
   Brightness (G.E.): 72
   Oil Absorption (GCOA), wt%: 180
   *Effective Density g/cc: 2.2
   Hegman Grind: 4
   MPD (micron): 8
183:
   Brightness (G.E.): 70
   Oil Absorption (GCOA), wt%: 175
   Hegman Grind: 3.5/MPD (micron): 8.9
   * In polyester resin

Grefco Minerals, Inc.: DIATOMITE Functional Fillers (Continued):

Processed Ditomite Functional Fillers:
WF:
  Color: White
  Brightness (G.E.): 90
  Oil Absorption (GCOA), wt%: 130
  Hegman Grind: 5.5
  MPD (Micron): 6.9
WFAB:
  Color: White
  Brightness (G.E.): 90
  Oil Absorption (GCOA), wt%: 130
  Hegman Grind: 4.5
  MPD (Micron): 8.25
395:
  Color: White
  Brightness (G.E.): 90
  Oil Absorption (GCOA), wt%: 130
  Hegman Grind: 4.5
  MPD (Micron): 7.7
WB-5:
  Color: White
  Brightness (G.E.): 88
  Oil Absorption (GCOA), wt%: 140
  Hegman Grind: 4.0
  MPD (Micron): 11.5
L-5:
  Color: White
  Brightness (G.E.): 87
  Oil Absorption (GCOA), wt%: 150
  Hegman Grind: 3.0
  MPD (Micron): 13.8
L-10:
  Color: White
  Brightness (G.E.): 86
  Oil Absorption (GCOA), wt%: 160
  MPD (Micron): 16.9
SP-5:
  Color: White
  Brightness (G.E.): 87
  Oil Absorption (GCOA), wt%: 160
  MPD (Micron): 24.6
PS:
  Color: Pink
  Oil Absorption (GCOA), wt%: 160
  Hegman Grind: 4.0
  MPD (Micron): 6.9
SF-5:
  Color: Pink
  Oil Absorption (GCOA), wt%: 175
  MPD (Micron): 15.4

Grefco Minerals, Inc.: DICAFLOCK Functional Fillers:

Dicaflock is powdered cellulose obtained from mechanically disintegrated cellulose prepared by processing bleached or unbleached cellulose obtained as a pulp from such fibrous materials as wood or cotton. It occurs as a white, odorless substance and consists of fibrous particles that disperse rapidly in water.

Dicaflock is available in various grades, exhibiting degrees of fineness ranging from a dense, free-flowing powder to a coarse, fluffy nonflowing material. It is insoluble in water, in dilute acids and in nearly all organic solvents. It is slightly soluble in sodium hydroxide TS. When bone dry, the white or bleached grades are at least 99.5% pure cellulose.

Dicaflock is an excellent filler for a wide variety of rubber and plastic products. Dicaflock improves dimensional stability, reduces green shrinkage, improves impact strength and improves drying rate of stable foams.

Applications include a wide variety of thermoset resins for injection molding and rubber compounds ranging from floor tiles to shoe soles. Dicaflock can be used as a plasticizer, bulking agent, arc intensifier, absorbent and lubricant.

Grades:
DF5:
    Fiber Length: Microns: 450
    Relative Flowrate: 2455
    Wet Cake Density: lbs/cu ft: 10.2

DF10:
    Fiber Length: Microns: 280
    Relative Flowrate: 1682
    Wet Cake Density: lbs/cu ft: 10.3

DF40:
    Fiber Length: Microns: 180
    Relative Flowrate: 1230
    Wet Cake Density: lbs/cu ft: 14.2

DF100:
    Fiber Length: Microns: 100
    Relative Flowrate: 880
    Wet Cake Density: lbs/cu ft: 15.2

DF200:
    Fiber Length: Microns: 50
    Relative Flowrate: 794
    Wet Cake Density: lbs/cu ft: 17.6

Grefco Minerals, Inc.: DICAPERL Hollow Glass Microspheres:

Dicaperl is a family of lightweight, hollow glass bubble fillers produced and marketed worldwide by Grefco Minerals. Available in a variety of particle size ranges, with or without surface modifications, Dicaperl is a very cost effective density reducing filler for any resin/binder system.

Silane-modified Dicaperl functional fillers substantially reduce weight, reduce shrinkage, improve impact, nailing, stapling, and sanding properties of molded parts. The finest grades are used in RP/Composites including glass fiber reinforced products formed by both spray-up and hand lay-up production processes, where surface detail is critical. The coarser grades are used in cast or molded parts not requiring ultra-fine surface appearance.

Fine particle size, unmodified grades of Dicaperl are quite effective as thickening, and anti-sag additives in specialty coatings. The coarser grades are used to produce texture and acoustical coating mixes, block filler paints, etc.

Dicaperl products are available in a range of densities, degrees of whiteness and strengths to suit most formulating requirements. Two proprietary coating processes are used, resulting in two distinct series of products, the standard "10 Series" and the high performance "20 Series."

HP-110:
    Particle Shape: Bubble
    Average Particle Size (Micron): 310
    *Effective Density: lb/ft3: 11.0

HP-210:
    Particle Shape: Bubble
    Average Particle Size (Micron): 110
    *Effective Density: lb/ft3: 11.5

HP-510:
    Particle Shape: Bubble
    Average Particle Size (Micron): 70
    *Effective Density: lb/ft3: 14.0

HP-710:
    Particle Shape: Bubble
    Average Particle Size (Micron): 65
    *Effective Density: lb/ft3: 14.7

HP-910:
    Particle Shape: Bubble
    Average Particle Size (Micron): 50
    *Effective Density: lb/ft3: 15

    * in polyester resin

Grefco Minerals, Inc.: PERLITE Functional Fillers:

Dicalite Perlite funtional fillers are a versatile family of mineral-based silicate products available throughout the world. These fillers are processed from an amorphous, naturally occurring glass found where there has been volcanic activity. When processed under the proper conditions the solid glass particles expand like popcorn up to 20 times their original volume. A specialized Grefco process incorporating expansion, selective milling, and sizing determines the particle shape of each filler product.

A variety of flake or di-, tri-, and tetrahedral particle shapes are obtained by this exclusive process. These unique shapes provide numerous functions including high surface area, permeability and reinforcement and have densities among the lowest of any of the mineral fillers. In addition to superior spatial loading, this property affords customers the added bonus of density control in their finished products. Dicalite Perlite fillers are off-white and do not interfere with the coloring of the products into which they are formulated.

Dicalite Perlite fillers are inert to all but strong acids or alkalis and have no residual surface reactivity. After processing they are free of any organic contamination.

Color: Off-White
Brightness (G.E.): 74
MOHS Hardness: 5.5
Moisture, %: 1.0
Ig. Loss (dry basis), %: 1.5
pH (10% slurry): 5.5-8.5

FF16:
Oil Absorption (GCOA), wt%: 210
*Effective Density, g/cc: 2.3
MPD (Micron): 11

FF26:
Oil Absorption (GCOA), wt%: 220
*Effective Density, g/cc: 2.2
MPD (Micron): 14

FF36:
Oil Absorption (GCOA), wt%: 240
*Effective Density, g/cc: 2.2
MPD (Micron): 24

FF56:
Oil Absorption (GCOA), wt%: 240
*Effective Density, g/cc: 1.2
MPD (Micron): 31

FF76:
Oil Absorption (GCOA), wt%: 240
*Effective Density, g/cc: 1.3
MPD (Micron): 37
* in polyester resin

J.M.Huber Corp.: Calcium Carbonate: Thermosets Product Selection Guide

Calcium Carbonate ranks on a volume basis second only to kaolin as the most widely used industrial mineral. The widespread use of ground calcium carbonate is due to a desirable combination of economic and physical characteristics such as availability, low cost, good color, low oil absorption and a wide range of particle sizes. Look to the W Series for lowest concentration of crystalline silica and no warning labels.

MARBLE ELITE Alpha:
  Cast Polymer: Gel Coated
  Composites: Electrical Laminates
W-3:
  Composites: Electrical Laminates/General Molded Products/
    Polyurethane Elastomer/PreForm/Pultrusion/RTM (Resin
    Transfer Molding)/SMC/BMC
W-4:
  Cast Polymer: Encapsulating/Potting
  Composites: Electrical Laminates/Filament Winding/General
    Molded Products/PreForm/Pultrusion/RTM (Resin Transfer
    Molding)/SMC/BMC
W-3N:
  Composites: General Molded Products/Polyurethane Elastomer/
    PreForm/Pultrusion/RTM (Resin Transfer Molding)
Q-325:
  Cast Polymer: Encapsulating/Potting
  Composites: Continuous Panel/Filament Winding/RTM (Resin
    Transfer Molding)/SMC/BMC/Spray-Up/Hand Lay-up
Q-200:
  Composites: Continuous Panel/Spray-Up/Hand Lay-Up
G-325:
  Cast Polymer: Encapsulating/Polymer
  Composites: Filament Winding/RTM (Resin Transfer Molding)/
    Spray-Up/Hand Lay-Up
G-260:
  Composites: Filament Winding/General Molded Products/Spray-
    Up/Hand Lay-Up
FLOMAX:
  Cast Polymer: Encapsulating/Polymer
  Composites: General Molded Products/RTM (Resin Transfer
    Molding)/SMC/BMC
S-325:
  Cast Polymer: Encapsulating/Potting
  Composites: Filament Winding/Spray-Up/Hand Lay-Up
S-200:
  Composites: Spray-Up/Hand Lay-Up

J.M. Huber Corp.: Kaolin Clays: Thermosets Product Selection Guide (Continued):

POLYFIL 8039:
Composites: Filament Winding, General Molded Products, Phenolic Molding, PreForm, Pultrusion, RTM (Resin Transfer Molding), SMC/BMC

Polyfil F:
Composites: Filament Winding, General Molded Products, PreForm, Pultrusion, RTM (Resin Transfer Molding), SMC/BMC

Polyfil 90:
Composites: Pultrusion, RTM (Resin Transfer Molding)

Intercorp Inc.: Diatomaceous Earth (Partial Grade Listing):

96C:
   Typical Loose Bulk Density Lbs/cu ft: 11
   Typical Sieve Analysis % Passing 325 Mesh: 96.0+-5%
   Typical Oil Absorption % by Weight: 142
   Typical Water Absorption %: 184

94C:
   Typical Loose Bulk Density Lbs/cu ft: 11
   Typical Sieve Analysis % Passing 325 Mesh: 94.0+-5%
   Typical Oil Absorption % by Weight: 142
   Typical Water Absorption %: 184

83A:
   Typical Loose Bulk Density Lbs/cu ft: 11
   Typical Sieve Analysis % Passing 325 Mesh: 83.0+-5%
   Typical Oil Absorption % by Weight: 133-151
   Typical Water Absorption %: 184

Applications:
   Thermosets, Thermoplastics, rubber, paper, fireproofing
boards, plasters, cements, stucco, gunites, paints, flow aid,
polish etc.

Intercorp Inc.: Sepiolite Clay-AEM (Preliminary-Partial Grade Listing):

WS3:
    Typical Loose Bulk Density Lbs/cu: 12
    Typical Sieve Analysis: Various particle sizes
    Typical Oil Absorption %: 95
    Typical Water Absorption %: 110
    Grade Note: 50% (+-5%) sepiolite content

LB-BG:
    Typical Sieve Analysis: Various particle sizes
    Typical Oil Absorption %: 220
    Typical Water Absorption %: 434
    Grade Note: 80% (+-5%) sepiolite content
    Specific Gravity: 1.03

LB-BR:
    Typical Sieve Analysis: Various particle sizes
    Typical Oil Absorption %: 120
    Typical Water Absorption %: 160
    Grade Note: 80% (+-5%) sepiolite content
    Specific Gravity: 1.04

Applications:
    Extender, Absorbent, Catalyst carrier, Suspension aid, Thixotropic agent, Soil conditioner, Binder

Intercorp Inc.: Wollastonite-Natural-KEMOLIT & TREMIN (Partial Grade Listing):

D1:
   Typical Aspect Ratio: 5:1
   Typical Particle Size Analysis: 5%-75u
   Grade Note: Coarse grade

M60:
   Typical Aspect Ratio: 15:1
   Typical Particle Size Analysis: 75%-75u
   Grade Note: Coarse, high aspect ratio

A60:
   Typical Aspect Ratio: 15:1
   Typical Particle Size Analysis: 82%-75u
   Grade Note: High aspect ratio (HAR)

A62:
   Typical Aspect Ratio: 20:1
   Typical Particle Size Analysis: 95%-75u
   Grade Note: High aspect ratio, high bulking

A64:
   Typical Aspect Ratio: 20:1
   Typical Particle Size Analysis: 96.5%-75u
   Grade Note: High aspect ratio, high bulking

S-300:
   Typical Aspect Ratio: 5:1
   Typical Particle Size Analysis: 99%-50u
   Grade Note: Fine mesh

S-400:
   Typical Aspect Ratio: 5:1
   Typical Particle Size Analysis: 99%-38u
   Grade Note: Fine mesh

S-500:
   Typical Aspect Ratio: 4:1
   Typical Particle Size Analysis: 99%-25u
   Grade Note: Fine mesh

VP-939 100 PST:
   Typical Aspect Ratio: 20:1
   Typical Particle Size Analysis: 44%-10u
   Grade Note: HAR, highly uniform

Intercorp Inc.: Wollastonite-Natural-KEMOLIT & TREMIN (Partial Grade Listing)(Continued):

VP-939-955 PST:
Typical Aspect Ratio: 20:1
Typical Particle Size Analysis: 44%-10u
Grade Note: HAR; highly uniform

VP-939-300 PST:
Typical Aspect Ratio: 20:1
Typical Particle Size Analysis: 60%-10u
Grade Note: HAR; highly uniform

WOLKRON 1025:
Typical Aspect Ratio: 20:1
Typical Particle Size Analysis: 63%-10u
Grade Note: HAR; highly uniform

VP-939-600 PST:
Typical Aspect Ratio: 20:1
Typical Particle Size Analysis: 82%-10u
Grade Note: HAR; highly uniform

Wolkron 1010:
Typical Aspect Ratio: 13:1
Typical Particle Size Analysis: 97%-10u
Grade Note: Ultra fine

General Applications:
Flux, paints & coatings, structural clay, friction, glass, ceramics, pigments, thermosets and thermoplastics.

Intercorp Inc.: Wollastonite-Surface Modified-FILLEX (Partial Grade Listing):

Selected FILLEX Grades:

1-AF1:
Typical Aspect Ratio: 15:1
Typical Particle Size Analysis: 82%-75u
Grade Notes: High aspect ratio (HAR)

2-AF1:
Typical Aspect Ratio: 15:1
Typical Particle Size Analysis: 92%-75u
Grade Notes: High aspect ratio

17-AF1:
Typical Aspect Ratio: 20:1
Typical Particle Size Analysis: 95%-75u
Grade Notes: High aspect ratio

6-AF1:
Typical Aspect Ratio: 5:1
Typical Particle Size Analysis: 99%-38u
Grade Notes: Fine mesh for high impact

9-AF1:
Typical Aspect Ratio: 3:1
Typical Particle Size Analysis: 97%-10u
Grade Notes: Ultra fine mesh for highest impact

VP-939-100 USST:
Typical Aspect Ratio: 20:1
Typical Particle Size Analysis: 44%-10u
Grade Notes: HAR; High DOI & elongation

VP-939-600 AST:
Typical Aspect Ratio: 20:1
Typical Particle Size Analysis: 82%-10u
Grade Notes: HAR; highest DOI for RRIM

General Applications:
Eng. resins-nylon, LCP, PBT, PC, PE, PP, olefins, urethanes & thermosets-polyurea, polyurethane, phenolic, polyester, epoxy. Kemolit grades are surface modified for specific customer applications.

Kaopolite Inc.: FERROSIL 14 Ferroaluminum Silicate:

Ferrosil 14 ferroaluminum silicate is a naturally occurring mineral mixture of almandite and pyrope. It has high uniform specific gravity, low oil absorption, a fine particle size, and high Mohs hardness.

Typical Physical Properties:
   Color: Light Buff
   Refractive Index: 1.83
   Specific Gravity: 4.0
   Density, Lbs/Solid Gallon: 33.32
   Bulking Value, Gallon/Lb.: 0.030
   Hardness Index (Mohs' Scale): 8-9
   pH (20% Solids): 8.5-9.0
   Oil Absorption (%, Gardner-Coleman): 18.0
   Moisture (% Max.): 1
   Median Particle Size (Microns): 3.8
   % Finer than 14 Microns: 98

Typical Chemical Analysis (%):
   SiO2: 41.2
   FeO + Fe2O3: 22.2
   Al2O3: 20.4
   MgO: 12.4
   CaO: 3.0
   Trace Elements: 0.8

Suggested Applications:
   Extender pigment for primers and other coatings, where hardness and high specific gravity are desired.
   Non-barium, high specific gravity alternative.
   Filler for abrasion resistant, plastic systems.

Kaopolite Inc.: Kaolin USP:

Kaolin USP hydrated aluminum silicate is a specially processed fine particle size kaolin designed specifically for use in formulations where high purity, bacteria control, and good color are desired.

Kaolin USP meets current United States Pharmacopoeia regulations and is ideal for use in cosmetic and pharmaceutical applications.

Typical Physical Properties:
G.E. Brightness (%): 88.5
Median Particle Size, Micron: 0.6
Moisture (Max.%): 1.0
Oil Absorption (%, Gardner-Coleman): 40
pH (20% Solids): 4.2-5.2
Wet Screen Residue (Max.% +325 Mesh): 0.03
Refractive Index: 1.56
Density, Lbs./Solid Gallon: 21.66
Bulking Value, Gallons/Lb.: 0.046
Specific Gravity: 2.58
Hardness Index (Mohs' Scale): 2

Kaopolite Inc.: KAOPOLITE Anhydrous Aluminum Silicates:

Kaopolite SF

Kaopolite SF is a fine, dry powder that is inert, insoluble, non-hygroscopic, and will not deteriorate in dry storage. Its fine particle size promotes good suspension in most liquid systems, and allows it to pass easily through aerosol valves. When incorporated into a product, Kaopolite SF's closely controlled water and oil demand properties yield consistent and economical compounds.

Kaopolite SF exhibits low absorptivity and is compatible with amine-functional silicones. When combined with harsher abrasives, Kaopolite SF shows a pronounced and beneficial "gentling" effect.

G.E. Brightness (%): 90
Median Particle Size, Micron: 0.7
Suggested Applications:
   Auto, metal, plastic, and household polishes
   Anti-block agent for plastic film
   Fine lapping compounds

Kaopolite 1152:

Kaopolite 1152 anhydrous aluminum silicate imparts very mild polishing properties. It has a fine particle size and disperses easily in both aqueous and non-aqueous systems. Kaopolite 1152 has the least amount of polishing action, which makes it ideal for use in clear coat auto finish polishes.

G.E. Brightness (%): 93
Median Particle Size, Micron: 0.8
Suggested Applications:
   Auto and plastic polishes
   A very "gentle" abrasive that speeds cleaning
   Anti-block agent for plastic film

Kaopolite 1168:

Kaopolite 1168 anhydrous aluminum silicate imparts aggressive polishing and cleaning properties. This product has a broad particle size distribution making it ideal for polishes which are used on heavily oxidized paint surfaces.

G.E. Brightness (%): 91
Median Particle Size, Micron: 1.8
Suggested Applications:
   Auto, metal and plastic polishes
   Anti-block agent for plastic film
   Aggressive cleaning compounds

Kaopolite Inc.: MAGOTEX Fused Magnesium Oxide:

Magotex fused magnesium oxide has the following unique combination of properties:
1. High thermal conductivity
2. Low oil absorption
3. Good electrical resistivity
4. Low moisture

Typical Physical Properties:
G.E. Brightness (%): 70.0
Median Particle Size (Microns): 14
Moisture (%): 0.1
Oil Absorption (% Gardner-Coleman): 17.0
pH (20% Solids): 11.0
Refractive Index: 1.73
Hardness (Mohs' Scale): 5.5-6.0
Specific Gravity: 3.6
Density, Lbs./Solid Gallon: 30.10
Bulking Value, Gallons/Lb.: 0.0332
Screen Residue (% Retained on 325 Mesh): 9.0
Resistivity (ohms-cm): 12,000

Thermal Properties:
Coefficient of Thermal Conductivity: 27 (BTU/(HR)(FT2)/(F/FT) at 140F. This compares to a value of 0.5-5.0 for many types of inorganic fillers, 118 for aluminum metal and 10 for aluminum oxide, at the same temperature.

Typical Chemical Analysis (%):
MgO: 95.0
SiO2: 3.0
CaO: 1.8
Fe2O3: 0.2

Suggested Application:
Filler for polymer and coating systems where high thermal conductivity with good electrical resistivity is required.

Kaopolite Inc.: SILTEX Fused Silica:

Siltex fused silica is a new, ultra-high performance extender pigment and filler for the coatings and polymer industries. This high purity fused silica is produced by melting selected glass sands at extremely high temperatures. The resulting amorphous product is then cooled, ground, and classified into two products, Siltex 44 and Siltex 44C. The unique amorphous structure of these products imparts improved optical, thermal, and physical properties to both coatings and polymer systems that cannot be obtained from naturally occurring silicas.

Siltex fused silica offers these unique advantages:
* Amorphous structure and ultra high purity imparts outstanding physical properties to many systems.
* Low refractive index permits formulating translucent coatings with outstanding optical properties.
* Chemical inertness and purity enhances chemical resistance.
* Ultra low magnetic iron contamination and amorphous structure give outstanding electrical properties to resin systems.
* Significantly lower thermal expansion than mineral products.
* Low moisture content and absorption enhances its use in coatings and moisture sensitive polymer systems.
* Extremely high specific resistance augments the electrical properties and dimensional stability of resin systems where moisture absorption and corrosion resistance are of prime importance.

Suggested Applications:
* Coating Systems:
  - High performance, corrosion-resistant coatings
  - Translucent or transparent coatings
  - Flat interior or exterior coatings
  - Marine coatings
  - Insulating electrical coatings
  - Moisture sensitive coatings
* Polymer Systems:
  - Low shrinkage and high dimensional stability in engineered thermoplastic systems
  - Excellent electrical properties in a wide range of polymer applications
  - Excellent for moisture cured urethanes and moisture sensitive CIP systems
* Cleaners and Polishes:
  - Hard metal, heavy duty liquid abrasive applications

Kaopolite Inc.: SILTEX Fused Silica (Continued):

Typical Physical Properties:
Siltex 44:
  G.E. Brightness (%): 90-92
  Moisture (% Max.): 0.1
  Oil Absorption (% Gardner-Coleman): 20-25
                (% Spatula Rub-Out): 27-33
  Median Particle Size (Microns): 8.8
  Screen Residue (Max. % +325 Mesh): 2.5
Siltex 44C:
  G.E. Brightness (%): 84-86
  Moisture (% Max.): 0.1
  Oil Absorption (% Gardner-Coleman): 18-23
                (% Spatula Rub-Out): 20-25
  Median Particle Size (Microns): 7.0
  Screen Residue (Max. % +325 Mesh): 5.0

General Physical Properties:
  pH (20% Solids): 6.5-7.5
  Refractive Index: 1.46
  Specific Resistance (ohms/cm): 150,000-200,000
  Specific Gravity: 2.17
  Density, Lbs/Solid Gallon: 18.08
  Bulking Value, Gallons/Lb.: 0.0553
  Hardness Index (Mohs' Scale): 5.5-6.0

Typical Chemical Analysis (%):
  SiO_2: 99.50 Minimum
  Al_2O_3: 0.20
  TiO_2: 0.03
  CaO: 0.01
  MgO: 0.01
  Fe_2O_3: 0.03
  Fe (Magnetic): 0.05
  Na_2O: 0.01
  K_2O: 0.01

Luzenac America, Inc.: Talc in Plastics: Products and Applications:

Luzenac America offers a broad selection of talc products designed specifically for plastic applications. Talc is a white, crystalline, platy mineral. It is the softest of all minerals, and in many applications, its shape imparts cost-effective and beneficial structural properties to the final product. Luzenac talcs are available in various particle size distributions and from a wide variety of Luzenac-owned ore sources.

Cimpact 699:
Median Particle Size (microns): 1.5
Color ("Y"): 87
Topsize (Hegman): 7.0//(microns): 10-15
Loose Bulk Density (lbs/ft3): 5-9
Typical Applications: High impact/automotive, TPO, nucleation

JetFil 700C:
Median Particle Size (microns): 1.5
Color ("Y"): 88
Topsize (Hegman): 7.0//(microns): 10-15
Loose Bulk Density (lbs/ft3): 40-50
Typical Applications: High impact/automotive

Cimpact 710:
Median Particle Size (microns): 1.8
Color ("Y"): 91
Topsize (Hegman): 7.0//Microns: 10-15
Loose Bulk Density (lbs/ft3): 6-10
Typical Applications: High impact/automotive, TPO, nucleation

Nicron 665:
Median Particle Size (microns): 1.5
Color ("Y"): 87
Topsize (Hegman): 6.5//Microns: 15-20
Loose Bulk Density (lbs/ft3): 6-10
Typical Applications: TPO, nucleation

Mistron ZSC:
Median Particle Size (microns): 2.1
Color ("Y"): 89
Topsize (Hegman): 6.0//Microns: 20-25
Loose Bulk Density (Lbs/ft3): 6-10
Typical Applications: Bubble nucleation, wire and cable

Mistron Monomix:
Median Particle Size (microns): 2.0
Color ("Y"): 89
Topsize (Hegman): 6.0//Microns: 20-25
Loose Bulk Density (lbs/ft3): 5-9
Typical Applications: Crystal nucleation, foamed polymers

JetFil 625C:
Median Particle Size (microns): 2.2
Color ("Y"): 88
Topsize (Hegman): 6.0//Microns: 20-25
Loose Bulk Density (lbs/ft3): 45-55
Typical Applications: Good impact/automotive copolymer

Luzenac America, Inc.: Talc in Plastics: Products and Applications (Continued):

Arctic Mist:
Median Particle Size (microns): 2.2
Color ("Y"): 88
Topsize (Hegman): 6.0//(Microns): 20-25
Typical Applications: Bubble nucleation, foamed plastics

Mistron Vapor R:
Median Particle Size (microns): 1.7
Color ("Y"): 87
Topsize (Hegman): 5.5//(Microns): 25-30
Typical Applications: Bubble nucleation, foamed plastics

JetFil 575C:
Median Particle Size (microns): 3.4
Color ("Y"): 87
Topsize (Hegman): 5.5//(Microns): 25-30
Typical Applications: Industrial/appliance, anti-block

Mistron 400C:
Median Particle Size (microns): 4.0
Color ("Y"): 89
Topsize (Hegman): 5.5//(Microns): 25-30
Typical Applications: Anti-block/film

Vertal 410:
Median Particle Size (microns): 7.5
Color ("Y"): 86
Topsize (Hegman): 4.0//(Microns): 45-55
Typical Applications: Industrial/automotive, better color

JetFil 350:
Median Particle Size (microns): 7.5
Color ("Y"): 84
Topsize (Hegman): 4.0//(Microns): 45-55
Typical Applications: Industrial homopolymer/copolymer

Stellar 420:
Median Particle Size (microns): 10.0
Color ("Y"): 88
Topsize (Hegman): 3.5//(Microns): 50-60
Typical Applications: Industrial, low cost homopolymer

TechFil 7599:
Median Particle Size (microns): 12.0
Color ("Y"): 77-80
Topsize (Hegman): 3.5//(Microns): 50-60
Typical Applications: Cost-effective appliance/automotive,
                      recycle

Vertal 97:
Median Particle Size (microns): 9-13
Color ("Y"): 74-78
Topsize (Hegman): 2.0//(Microns): 65-75
Typical Applications: Industrial, low cost homopolymer

Luzenac America, Inc.: Talc in Polypropylene

Luzenac America offers a broad line of talc products for use in polypropylene homopolymer. Talc increases the stiffness and heat distortion temperature (HDT) of polypropylene. It also reduces the coefficient of thermal expansion. Typical talc loadings are 20 to 40%. Because homopolymers do not have good impact strength, the talc products recommended are relatively coarse and vary based on price, talc content, color and long-term heat aging (LTHA).

Talc is a white, crystalline, platy mineral. It is the softest of all minerals. It must be mixed in the polymer melt to insure good dispersion.

Products and Applications:
Vertal 710:
   Median Particle Size (microns): 7.5
   Dry Color (GEB): 86
   Topsize: Hegman: 4.0//Microns: 45-55
   Loose Bulk Density (lbs/ft3): 20-24
   Typical Applications: Industrial/Automotive/Better Color
Jetfil 350:
   Median Particle Size (microns): 7.5
   Dry Color (GEB): 84
   Topsize: Hegman: 4.0//Microns: 45-55
   Loose Bulk Density (lbs/ft3): 22-26
   Typical Applications: Industrial/Homopolymer/Copolymer
Vertal 97:
   Median Particle Size (microns): 9-13
   Dry Color (GEB): 74-78
   Loose Bulk Density (lbs/ft3): 27-37
   Typical Applications: Industrial/Low Cost
Stellar 410:
   Median Particle Size (microns): 10
   Dry Color (GEB): 90
   Topsize: Hegman: 3.5//Microns: 50-60
   Loose Bulk Density (lbs/ft3): 20-24
   Typical Applications: Good LTHA/Better Color
Techfil 7599:
   Median Particle Size (microns): 12.0
   Dry Color (GEB): 77-80
   Topsize: Hegman: 3.5//Microns: 40-45
   Loose Bulk Density (lbs/ft3): 24-28
   Typical Applications: Cost-effective appliance/Automotive

Luzenac America, Inc.: Talc in PVC:

The substitution of talc for calcium carbonate in rigid PVC
systems results in:
* increased stiffness
* retention of tensile strength
* higher HDT
* lower coefficient of thermal expansion

Talc is a white, crystalline, platy mineral. Coarse talcs,
which are less costly than fine grinds, work well in non-impact
sensitive applications that require high stiffness and tensile
strength; for example, cooling tower plates. Other grinds are
suitable for swimming pool equipment-rails, ladders, filter
housings and pool liners-and credit card compounds where talc
improves printability while minimizing stress whitening due to
bending. Talc is also used cost-effectively in PVC films, PVC
pellet and cable dusting and PVC dispersions.

Talc reinforced thermoplastics yield higher heat distortion
temperatures and lower thermal expansion. These characteristics
indicate an extended service temperature range and improved
dimensional stability. In processing, talc builds torque which
promotes fusion in rigid vinyls. Talc also provides good hot
strength for downstream operations.

Cimpact 699:
Median Particle Size (microns): 1.2
Color (GEB): 87
Topsize (Hegman): 7.0//Topsize (microns): 10-15

Mistron Vapor R:
Median Particle Size (microns): 1.7
Color (GEB): 87
Topsize (Hegman): 5.5//Topsize (microns): 25-30

Arctic MIST:
Median Particle Size (microns): 2.2
Color (GEB): 88
Topsize (Hegman): 6.0//Topsize (microns): 20-25

JetFil 575P:
Median Particle Size (microns): 3.4
Color (GEB): 87
Topsize (Hegman): 5.5//Topsize (microns): 25-30

JetFil 500:
Median Particle Size (microns): 5.0
Color (GEB): 86
Topsize (Hegman): 5.0//Topsize (microns): 30-40

Vertal 410:
Median Particle Size (microns): 7.5
Color (GEB): 86
Topsize (Hegman): 4.0//Topsize (microns): 45-55

JetFil 350:
Median Particle Size (microns): 7.5
Color (GEB): 84
Topsize (Hegman): 4.0//Topsize (microns): 45-55

Malvern Minerals Co.: NOVACITE:

Origin: Novaculite Uplift--Arkansas, USA
Scientific Description: A natural microform of quartz in a
  bound and free state of subdivision

Typical Chemical Composition: Apparent Bulk Density (Approx.):
  SiO_2:   99.12%
  Fe_2O_3:   0.04   Loose Packed: lbs/ft3: 83
  TiO_2:   0.015   lbs/gal: 11
  CaO: 0.0   True Density: 22.07 lbs/gal
  MgO: 0.0   pH (typical): 6.3-7.0
  Al_2O_3:   0.61   Melting Point: 1500-1700C
  LOI: 0.20   Specific Heat: Mean between 0
                                   and 200C: 0.192 Cal/g/C
  Specific Gravity: 2.65
  Hardness: 7.0 Mohs
                   820 Knoop

Transmission For Radiation:
  Very transparent to the ultraviolet and the visible spectrum,
but opaque for the infrared beyond 7.0 microns (quartz).

Description of Particles:
  Less than 7 microns is singular, platey and lamellar. Greater
than 7 microns, clusters. (Bowen Edition American Journal
Science-1952-Page 244)
  Surface Area: 2 M2g BET Method
  Moisture Loss: 3 Hours @ 110C - None
  Loss on Ignition: 30 min. @ 1000C - 0.20%
  Silanol Count (SiOH): 4-6/100A2

Thermal Conductivity* - (Quartz)
              100F   200F   300F
c-Axis   6.40   5.40   5.02
a-Axis   3.40   3.00   2.60
   *BTU/(hr) (sq.ft.) (F/Ft)

Malvern Minerals Co.: NOVACITE (Continued):

1250 Novacite:
1250 Novacite is a premium 325 mesh product. Normally, it is 100% finer than 44u. This outstanding product has been preferred for over 25 years in thermoset molding compounds.
Typical Applications:

Fluidized bed coatings	Casting & potting resins
Polyurethane grouts	Molding compounds
Polycarbonate	Electrostatic coatings
Pipe linings	Industrial coatings
Abrasive medium (Wet blasting)	

DAPER Novacite:
Daper Novacite lies between 1250 and L-207A in fineness. It is the coarsest of the finer grades. Daper is a candidate as requirements become stricter with regard to fineness, gloss control and abrasivity. Even though it is quite fine, Daper still has the uniqueness of low binder demand with superior rheological flow properties.
Typical Applications:

Silicone rubber extender	Silicone rubber dusting powder
Electrostatic coating	Molding compounds
Fluidized bed coatings	Casting and potting compounds
Most all coatings and paints	

L-207A Novacite:
L-207A Novacite is the "star" of the Line. Normally the product will disperse to a 7 Hegman Grind. All particles are individually platey with few clusters. 98% of the particles will fall between one and ten microns. L-207A as fine as it is still has the uniqueness of low binder demand with superior flow properties.
Typical Applications:

Silicone rubber extender	Silicone rubber dusting powder
Electrostatic coating	Molding compounds
Fluidized bed coatings	Casting and potting compounds
Most all coatings and paints	
Light diffusing characteristic control	

L-337 Novacite:
L-337 Novacite is finer than L-207A Novacite. Normally the product will disperse to a 7 Hegman Grind. All particles are individually platey with few clusters. 98% of the particles will fall between one and ten microns. L-337 as fine as it is still has the uniqueness of low binder demand with superior flow properties.
Typical Applications:

Silicone rubber extender	Silicone rubber dusting powder
Electrostatic coating	Molding compounds
Fluidized bed coatings	Casting and potting compounds
Most all coatings and paints	
Light diffusing characteristic control	

Malvern Minerals Co.: NOVAKUP:

Novakup has you covered...if you're looking for:
* Reduction in expansion coefficients
* Adhesion to metal substrates
* Control of water vapor transmission
* Wet strength retention (Physical)
* Wet strength retention (Electrical)
* Increased chemical resistance
Novakup is surface treated Novacite
Cross linking is the reason why Novakup has you covered.
Novakup is a naturally occuring microform of silica which
can have the surface treated to provide an organofunctional
group ready on the surface. These groups* can be readily attached
to Novakup surfaces:
   *Epoxy, methacryloxy, vinyl, chloropropyl, amine, mercaptan,
   methyl, polyamine, tri-isostearate, styryl-amine, tri-
   lauryl myristyl, glass resin

Per Cent Finer than Sieve Opening:

200-R:
   Micron Diameter: 74/U.S. Sieve Series Number: 200: 98.12
   Avg Part Size: 11.5u

325-R:
   Micron Diameter: 37/U.S. Sieve Series Number: 400: 93.07
   Avg Part Size: 9.5u

625-R:
   Micron Diameter: 20/U.S. Sieve Series Number: 625: 81.10
   Avg Part Size: 8.5u

1250-R:
   Micron Diameter: 20/U.S. Sieve Series Number: 625: 92.00
   Avg Part Size: 6.5u

Daper-R:
   Micron Diameter: 15/U.S. Sieve Series Number: 950: 95.80
   Avg Part Size: 5.5u

L-207A-R:
   Micron Diameter: 10/U.S. Sieve Series Number: 1250: 100.00
   Avg Part Size: 4.0u

L-337-R:
   Micron Diameter: 10/U.S. Sieve Series Number: 1250: 100.00
   Avg Part Size: 3.45u

L-335-R:
   Micron Diameter: 10: U.S. Sieve Series Number: 1250: 100.00
   Avg Part Size: 2.40u

Mississippi Lime Co.: Precipitated Calcium Carbonates:

Mississippi HO-M60 Milled Precipitated Calcium Carbonate:
  Mississippi HO-M60 Milled is specifically designed as an
extender pigment for coatings. It provides excellent hide, good
blue-white shade and a high Hegman grind.

Typical Physical Analysis:
  Pounds/Gallon @ 25C: 23.7
  Bulking Value: One Pound Bulks: 0.0422
  Specific Gravity: 2.85
  Dry Brightness, G.E.: 97.5%
  Mean Particle Size-Sedigraph: 0.9 Micron
  Oil Absorption (Rub Out)-ASTM D-281: 50.0
  pH: 9.5
  BET Surface Area: 10.0 m2/g
  325 Mesh Residue: 0.01%
  Crystal Type: Acicular Aragonite
  Hardness: Mohs: 4.0
  Refractive Index: 1.68
  Apparent Dry Bulk Density-Loose: 14.0 lbs./ft3
  Apparent Dry Bulk Density-Packed: 27.0 lbs./ft3
  Hegman grind: 6.0
Typical Chemical Analysis: CaCO3: 98.60%

Mississippi M60 Spray Dried Precipitated Calcium Carbonate:
  Mississippi M60 Spray Dried is an outstanding filler pigment
for paper applications. It provides good opacity as well as
excellent retention and drainage properties.

Typical Physical Properties:
  Moisture: 0.7%
  Specific Gravity: 2.85
  Dry Brightness, G.E.: 97.0%
  Refractive Index: 1.68
  Mean Particle Size-Sedigraph: 0.95 Micron
  BET Surface Area: 9.0 m2/g
  Oil Absorption (Rub-Out)-ASTM D-281: 50.0
  pH: 9.5
  Einlehner Abrasion: 5.0
  325 Mesh Residue: 0.01
  Crystal Type: Acicular Aragonite
  Zeta Potential: +21.0
  Apparent Dry Bulk Density-Loose: 22 lbs/ft3
  Apparent Dry Bulk Density-Packed: 38 lbs/ft3
Typical Chemical Analysis: CaCO3: 98.60%

Maississippi Lime Co.: Precipitated Calcium Carbonates (Continued):

MAGNUM GLOSS Milled Precipitated Calcium Carbonate:
Magnum Gloss is designed to provide a superior level of gloss and brightness in satin and semi-gloss paints as well as industrial coatings. This significantly higher gloss is due in part to Magnum Gloss' tightly controlled particle range.

Typical Physical Analysis:
Form: Dry Powder
Bulking Value: One Pound Bulks: 0.0422
Bulk Density: Loose: 19.0 lbs/ft3
             Packed: 32.0 lbs/ft3
Pounds/Gallon: 23.7
Hegman Grind: 5.0
Specific Gravity: 2.85
Dry Brightness, G.E.: 98.0%
Refractive Index: 1.68
Mean Particle Size-Sedigraph: 0.38 Micron
Particle Size (% Finer than 2 Micron): 98.5%
Oil Absorption: 28.0
BET Surface Area: 12.0 m2/g
pH: 9.5
325 Mesh Residue: 0.01%
Crystal Type: Acicular Aragonite
Typical Chemical Analysis: CaCO3: 98.60%

Mississippi Lime Co.: Slurry Precipitated Calcium Carbonates:

Mississippi HO-M60 Slurry Precipitated Calcium Carbonate:
Mississippi HO-M60 is specifically designed for paper filling
applications. It provides high opacity, excellent retention,
good drainage, and bulking properties.

Typical Physical Analysis:
   Slurry Solids: 50.0%
   Weight/Gallon: 12.35
   Dry Pounds/Gallon: 6.18
   Specific Gravity: 2.85
   Dry Brightness, G.E.: 98.0%
   Refractive Index: 1.68
   Mean Particle Size-Sedigraph: 1.4 Micron
   BET Surface Area: 10.0 m2/g
   pH: 10.5
   Einlehner Abrasion: 5.0
   325 Mesh Residue: 0.008%
   Crystal Type: Acicular Aragonite
   Zeta Potential: +13.0
Typical Chemical Analysis: CaCO3: 98.60%

Stabilized HO-M60 Slurry Precipitated Calcium Carbonate:
HO-M60 is specially designed as a extender pigment for coatings. It provides very good hide and has excellent high brightness and blue-white shade which will result in clearer, bright-
er paint colors.

Typical Physical Analysis:
   Slurry Solids: 50%
   Weight/Gallon: 12.35
   Dry Pounds/Gallon: 6.18
   Specific Gravity: 2.85
   Dry Brightness, G.E.: 97.5
   Refractive Index: 1.68
   Mean Particle Size-Sedigraph: 1.3 Micron
   BET Surface Area: 10.0 m2/g
   pH: 9.5
   Einlehner Abrasion: 5.0
   325 Mesh Residue: 0.008%
   Crystal Type: Acicular Aragonite
   Oil Absorption (Rub Out)-ASTM D-281: 53.0
Typical Chemical Analysis: CaCO3: 98.60%

Mississippi Lime Co.: Slurry Precipitated Calcium Carbonates (Continued):

MAGNUM GLOSS 72% Precipitated Calcium Carbonate Slurry:
Mississippi Lime Co. has designed Magnum Gloss for use in glossy, high brightness coated paper and board applications. Due to its tightly controlled particle size range, it will provide significantly higher sheet gloss than other calcium carbonates. Another unique characteristic is its Hercules high shear rheology in the range of 2.0 dynes at 4400PM. Magnum Gloss can provide improvements in gloss, opacity, brightness, ink receptivity and runnability on the coater--all of which combine to make this a clearly exceptional product.

Typical Physical Analysis:
    Slurry Solids: 72.0%
    Weight/Gallon: 15.67
    Dry Pounds/Gallon: 11.28
    Specific Gravity: 2.85
    Dry Brightness, G.E.: 98.0%
    Refractive Index: 1.68
    Mean Particle Size-Sedigraph: 0.38 Micron
    Particle Size (% Finer than 2 Micron): 98.5%
    Oil Absorption: 28.0
    BET Surface Area: 12.0 m2/g
    pH: 9.5
    Einlehner Abrasion: 5.0
    325 Mesh Residue: 0.001%
    Crystal Type: Acicular Aragonite
    Hercules Viscosity: 2.0 Dynes at 4400 RPM
    Brookfield Viscosity: 500 cps @ 20 RPM
Typical Chemical Analysis: $CaCO_3$: 98.60%

Mississippi Lime Co.: Sluury Precipitated Calcium Carbonates (Continued):

Mississippi 70% M60 Slurry Precipitated Calcium Carbonate:
Mississippi 70% M60 Slurry is a versatile pigment for paper filling and coating applications. As a filler, it provides high brightness, good opacity, and excellent drainage properties. In coatings, it gives outstanding ink receptivity and high brightness properties.

Typical Physical Analysis:
Slurry Solids: 70.0%
Weight/Gallon: 15.28 lbs
Dry Solids/Gallon: 10.7 lbs
Specific Gravity: 2.85
Dry Brightness, G.E.: 98.0%
Refractive Index: 1.68
Mean Particle Size-Sedigraph: 1.1 Micron
BET Surface Area: 12.5 m2/g
Oil Absorption (Rub Out) ASTM D-281: 33.0
pH: 11.0
Einlehner Abrasion: 4.0
325 Mesh Residue: 0.05%
Crystal Type: Acicular Aragonite
Zeta Potential: -5.0
Typical Chemical Analysis: CaCO3: 98.60%

Precipitated Calcium Carbonate: Mississippi 72% M60 Slurry:
Typical Physical Analysis:
Slurry Solids: 72.0%
Weight/Gallon: 15.67 lbs
Dry Solids/Gallon: 11.28 lbs
Specific Gravity: 2.85
Dry Brightness, G.E.: 98.0%
Refractory Index: 1.68
Mean Particle Size-Sedigraph: 0.45 Micron
BET Surface Area: 11.4 m2/g
Oil Absorption (Rub Out) ASTM D-281: 30.0
pH: 9.5
Einlehner Abrasion: 4.0
325 Mesh Residue: 0.01%
Crystal Type: Acicular Aragonite
Brookfield Viscosity: 800 cps @ 20 RPM
Typical Chemical Analysis: CaCO3: 98.60%

NYCO: Chemically Modified Minerals:

The surface modification of minerals with coupling and/or
wetting agents changes a mineral from a utilitarian filler
to a functional component of a polymer composite...adding
performance values that the base resin alone does not possess.
Optimum performance of a mineral reinforced resin requires
proper interaction of the chemical treatment with the surface
of the mineral and the polymer. This requirement is best
served by a surface modified mineral, thus eliminating the
critical variable of assuring interaction with the mineral
surface and insuring the same properties batch after batch.
Key consideration in developing high performance mineral
reinforcements include the selection of:
   * base mineral or mineral blend
   * surface treatment chemical(s) and concentration level
   * technique(s) of applying/reacting the chemical to the
     mineral surface
Every mineral has its own set of properties: chemical composition...particle shape and size...surface area...physical,
thermal and electrical properties.
Equally critical as the mineral selection is the choice of
the surface modifying chemical(s). Selection of the surface
modifying chemical is directed by the polymer system to be
used, its processing and compounding limitations, required
composite properties, and mineral reinforcement used.
Typical Chemical Treatments:
Proprietary Coupling Agents: Nyco's broad experience in surface
treatment of minerals has resulted in the development of a
number of unique proprietary surface treatments.
Organic Treatments: This broad grouping of chemicals, typical
examples of which are listed below, exhibits a wide variation
of stability and reactivity in their neat form. Reactivity
considerations include sensitivity to temperature, light,
heat, moisture and the catalytic effects of trace contaminants,
all serious concerns in material storage, handling, and use.
Commercially available coupling agents/wetting agents
include:
   * Silane coupling agents
   * Organosilicon chemicals
   * Polymeric ester wetting agents/hydrophobes
   * Titanate wetting agents

Nyco: Chemically Modified Minerals (Continued):

Description:
  Characteristics of Surface Modified Minerals:
  The general benefits of chemically modified minerals are:
  * Improved mineral dispersion and wet out by the matrix
    resin leading to mineral deagglomeration and resistance
    to reagglomeration.
  * Improved processing
  * Improved mineral/polymer bonding
  * Upgraded mechanical properties
  * Improved electrical properties
  * Reduced water sensitivity of the mineral polymer interface
    resulting in maintenance of composite properties in wet
    environments.
  Why risk the uncertainty of in-process addition of chemicals
when surface modified minerals offer so many advantages:
  * Consistent compound properties and handling characteristics
    batch to batch.
  * Development of critical wet electrical and wet strength
    properties.
  * Lower melt viscosity resulting in better flow into the
    mold.
  * Chemical level adjustment is unnecessary from formulation
    to formulation which may be required in integral addition.
  * Reduced dependence on processing equipment design.
  * Reduced sensitivity to the order of addition of ingredients
    in a molding compound formulation.
  * Broader processing window.
  * Economics: Surface modified minerals are among the most
    cost-effective material investments you can make.

NYCO: Advantages of WOLLASTOCOAT Surface Modified Wollastonite:

Nylon:
* Low moisture absorption
* Industry standard for platable grades
* Cost-effective milled glass replacement
* Improved mold flow
* Uniform heat distribution
* Better dimensional stability
* Improved surface characteristics
* Higher heat-distortion temperature
* Upgraded mechanical properties
* Reduced wall thickness

Phenolic Molding Compounds:
* Low resin demand, better wetout
* High loadings - good flow
* Faster mold cycle time
* Highly reinforcing
* Cost-effective milled glass replacement
* Improved machinery of molded parts
* Low moisture absorption
* Enhanced compression strength
* Excellent dimensional stability
* Improved physical and electrical properties at elevated temperatures

Epoxy:
* Higher mineral loading
* Low moisture absorption
* Improved thermal conductivity
* Improved resistance to thermal shock
* Higher heat-distortion temperature
* Better machinability of molded parts
* Greater impact resistance
* Improved crack resistance
* Cost reduction

Polyester:
* Viscosity reduction
* Low moisture absorption
* Improved surface appearance
* Promotes better platability
* Excellent heat stability
* Improved flexural strength
* Better electrical insulating properties
* Cost effective partial glass replacement

Polyurethane:
* Cost-effective glass replacement in RRIM
* Better distinction of image in RRIM than glass fiber
* Low viscosity
* Nonsettling, no hardpanning in casting compounds
* Low moisture absorption
* Isotropic replacement in RRIM
* High flexural modulus
* Lower coefficient of thermal expansion
* Reduced heat sag

NYCO: NYAD Wollastonite Calcium Metasilicate:

Nyad G Wollastonite:
Description: High aspect ratio large median diameter grades.

G Wollastocoat:
Description: Surface modified wollastonite product line, using Nyad G as the base.

Nyad M1250 Wollastocoat:
Description: Fine particle size powder grade.

10 Wollastocoat/M1250 Wollastocoat:
Description: Surface modified wollastonite product line, using Nyad 1250 as the base.

Nyad M400 Mesh Wollastonite:
Description: Base grade for 400 Wollastocoat surface modified wollastonite product line.

Nyad 325:
Description: Varying mesh size powder grade.

Nyad 200:
Description: Varying mesh size powder grade.

Description (All of the above):
Wollastonite is a naturally occurring calcium metasilicate. It is the only commercially available pure white mineral that is wholly acicular; typical length to diameter ratios range from 3:1 to 20:1. The world's largest known and purest deposit is owned and operated by Nyco.

NYCO: NYGLOS Wollastonite:

NYGLOS 4, 5, 8, 12, 20, M3, M15, M20

High aspect ratio fine median diameter grades. Wollastonite is a naturally occurring calcium metasilicate. It is the only commercially available pure white mineral that is wholly acicular; typical length to diameter ratios range from 3:1 to 20:1.

By special milling techniques, Nyco is able to retain the acicular integrity of the Wollastonite and produce many high aspect ratio grades.

The world's largest known and purest deposit (in Mexico) is operated by Nyco, with another high quality deposit in New York State.

New York deposit beneficiated in a dry process using high intensity magnetic separators. Mexican deposit beneficiated in a wet process using flotation.

Other high aspect ratio, fine median diameter grades available:

Ultrafine I & II
Primglos I & II

Chemical Composition:

Nyad wollastonite is the purest known deposit in the world
Calcium Oxide: 47.0%+
Silica: 50.0%+
Other Components: Remainder

NYCO: Chemically Modified FLAKEGLAS -1/8", 1/64":

Fiberglas Flakeglas reinforcements, produced by Owens Corning Fiberglas, are thin glass flakes that have been hammermilled through a designated screen size. These flakes are manufactured from chemical resistant "C" glass. There are two products, 1/8" and 1/64". These flakes have extremely high stiffness at low loading levels versus other reinforcements. Unlike micaceous products, the Flakeglas reinforcement does not delaminate under stress.

NYCO applies a chemical surface treatment, through a proprietary process, to these Flakeglas reinforcements. The resultant chemically modified Flakeglas exhibits:

* improved flow characteristics
* less air entrapment
* improved wet-out and dispersion
* improved strength properties
* low permeability to moisture and solvents
* improved surface

Physical Properties (Unmodified):
Screen Analysis - 1/8":
    Screen Number: 6
    Nomonal: 0

    Screen Number: 12
    Nominal: trace

    Screen Number: 20
    Nominal: 9.1

    Screen Number: 40
    Nominal: 33.3

    Screen Number: 70
    Nominal: 33.8

    Screen Number: pan
    Nominal: 20.5

Screen Analysis: 1/64":
    Screen Number: 40
    Nominal: 0.5

    Screen Number: 70
    Nominal: 21.0

    Screen Number: pan
    Nominal: 76.7

Ohio Lime, Inc.: OHSO Dolomitic Limestone:

Ohso Pulverized Dolomitic Limestone is quarried, crushed, kiln dried and sized. Its size, purity and flowability make it attractive for several industrial applications. Its low moisture content minimizes storage and handling problems during cold weather.

Filler: Ohso Pulverized Dolomitic Limestone may be used as a filler in rubber, plastics, ceramics, caulking compounds, adhesives, sealants, roof coating and texture paint. Anywhere there is a need for an inexpensive and relatively inert filler, Ohso should be considered.

Agricultural: Ohso Pulverized Dolomitic Limestone is used to neutralize acid soils on farms, in gardens and on lawns. The ability to control pH makes it an important ingredient in silage or anywhere methane gas evolution is a problem.

Other Uses: Ohso has application as a welding flux ingredient and as a pre-coating on baghouse dust collector bags.

Typical Chemical Analysis:
    Calcium Carbonate: 54.23%
    Magnesium Carbonate: 44.80%

In Terms of an Oxide:
    Calcium Oxide: 30.28%
    Magnesium Oxide: 21.28%
    Iron Oxide: 0.069%
    Sulfur: 0.011%
    Loss on Ignition: 47.94%
    Moisture: 0.05%

Typical Screen Analysis (Accumulative):
Mesh	%
20 Mesh:	0.0%
40 Mesh:	0.0%
100 Mesh:	15.1%
200 Mesh:	41.5%
-200 Mesh:	58.5%

GE Brightness: 52.83
Bulk Density: 90.2 lbs/ft3 (1445 kg/m3)
Neutralizing Power in Terms of CaCO3: 106

Ohio Lime, Inc.: STONELITE Dolomitic Limestone:

Stonelite Pulverized Dolomitic Limestone is quarried, crushed, kiln dried and sized. Its size, purity and flowability make it attractive for several industrial applications. Its low moisture content minimizes storage and handling problems during cold weather.

Filler: Stonelite Pulverized Dolomitic Limestone may be used as a filler in rubber, plastics, ceramics, caulking compounds, adhesives, sealants, roof coating and latex or texture paint. Anywhere there is a need for an inexpensive and relatively inert filler where fineness is critical, Stonelite should be considered.

Filler Coating: Stonelite also has a specialized application as a precoating for baghouse dust collection bags.

Typical Chemical Analysis:
    Calcium Carbonate: 54.23%
    Magnesium Carbonate: 44.80%

In Terms of an Oxide:
    Calcium Oxide: 30.28%
    Magnesium Oxide: 21.28%
    Iron Oxide: 0.069%
    Sulfur: 0.011%
    Loss on Ignition: 47.94%
    Moisture: 0.05%

Typical Screen Analysis (Accumulative):
     60 Mesh: 0.0%
    200 Mesh: 0.0%
    325 Mesh: 6.3%
   -325 Mesh: 93.7%

GE Brightness: 79.0
Bulk Density: 68.2 lbs/ft3 (1092 Kg/m3)
Neutralizing Power in Terms of CaCO3: 106

Perlite Institute, Inc.: Perlite as an Ultrafine Filler:
(Courtesy: Whittemore Co., Inc.):

Why Perlite Ultrafine Fillers?
Several important characteristics of ultrafine perlite make
it extremely useful as a filler:
Very light weight
Reinforcing structure
Fine particle size
High brightness
Mild abrasiveness
Lower volumetric cost
Availability

Typical Product Applications:
Perlite ultrafine fillers have been used in a number of
proprietary applications for many years. Some of these
application areas include:

silicone rubber	thermosetting resins
synthetic rubbers	polyester
natural rubber	others
thermoplastic resins	paint
polyethylene	coatings
PVC	polishes
others	cleansers

Typical Chemical Analysis:*
Silicon: 33.8
Aluminum: 7.2
Potassium: 3.5
Sodium: 3.4
Oxygen (by difference): 47.5
Bound Water: 3.0
* All analysis are shown in elemental form

Typical Product Data:
Color: White
G.E. Brightness, %: 70-80
Refractive Index: 1.47
Specific Gravity: 2.2-2.4
Apparent or Bulk Density, lb/ft3: 5-15
                        gm/cc: 0.08-0.24
pH: neutral
Oil Absorption: 120-240
Softening Point, F: 1800
                 C: 980
Moisture, %: <1.0
Water Absorption: 195-350
Ignition Loss, 3 hr 1700F (930C): 1.5% max
Mean Particle Diameter, Microns: as small as 10

Perlite Institute, Inc.: Perlite Lightweight Hollow Spheres: (Courtesy: Whittemore Co., Inc.):

Advantages of Perlite Hollow Spheres or Bubbles as Fillers:
Because of their unique multicellular structure, these lightweight perlite fillers can provide many advantages:
1. Lightweight perlite fillers are effective bulk fillers because of their low density.
2. Lightweight perlite fillers exhibit a uniform white color that has minimal to no effect on the color of the finished product.
3. The particle shape of lightweight perlite fillers promotes good bonding between the perlite and resins.
4. Some properties such as impact resistance and tensile stength may be enhanced by the particle shape and size of lightweight perlite fillers.
5. Lightweight perlite fillers can reduce raw material costs while improving workability.
6. When used with other fire retardant materials, lightweight, inert and inorganic perlite fillers can enhance fire resistance.
7. Perlite fillers are generally inert and perlite is listed in the U.S. Food Chemicals Codex.

Because of these characteristics, perlite lightweight hollow spheres or bubbles have been found to be cost effective in a number of applications including the manufacture of adhesives, auto body putty, cultured marble, coatings, patching compounds and stucco

Typical Chemical Analysis*:
Silicon: 33.8
Aluminum: 7.2
Potassium: 3.5
Sodium: 3.4
Oxygen (by difference): 47.5
Bound Water: 3.0
* All analysis are shown in elemental form

Typical Product Data:
Bulk Density, lb/ft3: 4-10
              kg/m3: 64-144
Alkalinity: 0.030-0.035
Oil Absorption: 175-350*
Thermal Conductivity, Btu-in/h-ft2-F: 0.30-0.40
                     W/m-K: 0.050-0.057
Surface pH: Neutral
Color: White
Average Particle Size, Microns: 40-310
Effective Density, lb/ft3: as low as 12*
                 kg/m3: as low as 192*
* Varies with product and application

Pierce & Stevens Corp.: DUALITE Microspheres:

I. Dualite's lightweight, hollow microspheres:
   * lower volume costs
   * reduce weight
   * lower VOC emissions

II. Dualite's resilient, spherically shaped microspheres:
   * uniformly distribute mechanical stress

III.Dualite's elastic microsphere shells:
   * non-friable (pressure stable)
   * shear-stable (mixable at high speeds)

IV.Dualite's coated microspheres:
   * enhance formulations compatibility
   * improve handling characteristics

V. Dualite's narrow particle size distribution:
   * improves workability/machinability
   * reduces shrinkage

Dualite Grade:
MS7020:
   Shell Polymer: ACN
   Density (grams/cc): 0.135+-0.015
   Particle Size Mode (Peak): 20-40u
   Solvent Resistance: Better
   Heat Resistance: 300F+

M6001AE:
   Shell Polymer: PVDC
   Density (grams/cc): 0.130+-0.02
   Particle Size Mode (Peak): 40-60u
   Solvent Resistance: Good
   Heat Resistance: 250F

M6050AE:
   Shell Polymer: ACN
   Density (grams/cc): 0.130+-0.03
   Particle Size Mode (Peak): 90-110u
   Solvent Resistance: Best
   Heat Resistance: 350F+

MS7000:
   Shell Polymer: ACN
   Density (grams/cc): 0.065+-0.005
   Particle Size Mode (Peak): 130-150u
   Solvent Resistance: Better
   Heat Resistance: 325F+

Pierce & Stevens Corp.: MICROPEARL Microspheres:

Micropearl products are heat expandable polymeric microspheres that possess a wide range of performance characteristics
with applicability to a broad spectrum of end uses.

Features:
* Expandability over a wide range of temperatures
* Wide scope of performance characteristics
* Available as a dry powder or wet cake

Typical End-Uses:
* PVC foam   * Unsaturated polyester resins
* Expandable printing ink   * Moldable plastics
* Putty & synthetic wood   * Paint
* Automotive underbody coatings   * Rubber

All Micropearl Grades available in wet or dry form:
No Suffix: -30% H2O   Suffix D: <5% H2O

F-30/F-30D:
Expansion Starting Temp (C): 95
Optimum Expansion Temp (C): 120-140
Heat Resistance: Good
Solvent Resistance: Good
Suggested Applications: Puff Ink Non-Woven

F-46/F-46D1:
Expansion Starting Temp (C): 110
Optimum Expansion Temp (C): 130-150
Heat Resistance: Good
Solvent Resistance: Very Good
Suggested Applications: Puff Ink Wall Paper

F-80S/F-80SD1:
Expansion Starting Temp (C): 140
Optimum Expansion Temp (C): 160-170
Heat Resistance: Very Good
Solvent Resistance: Excellent
Suggested Applications: Anti-skid

F100/F100D1:
Expansion Starting Temp (C): 150
Optimum Expansion Temp (C): 180-190
Heat Resistance: Excellent
Solvent Resistance: Excellent
Suggested Applications: Plastic Molding

Piqua Minerals: PIQUA Minerals Fillers:

Typical Chemical Analysis:
   Calcium Carbonate ($CaCO_3$): 85-88%
   Magnesium Carbonate ($MgCO_3$): 10-13%
   Silicon Dioxide (SiO_2): <1%
   Aluminum Oxide (Al_2O_3): <0.5%
   Iron Oxide (Fe_2O_3): <0.5%
   Sulfur (S total): <0.1%
   Manganese Oxide (MnO): <0.04%
   Moisture: <0.2%
   Loss on Ignition (LOI): 43.5

Piqua Minerals Filler 60:
   Specific Gravity (gm/cc): 2.71
   Median Particle Dia. (microns, Sedigraph): 6
   Wt. per Gallon (lbs, solid): 22.50
   pH of Saturated Solution: 9.40
   Index of Refraction: 1.60
   Dry Brightness (Hunter): 80-82
   Oil Absorption (gm/100 gm): 16-20
   Mohs Hardness: 3
   Bulk Density (lbs/cu ft, loose, approx.): 34
   Free Moisture @ 110C: 0.20 max.
   Percent Passing 325 Mesh Screen: 99.5
Typical Chemistry:
   Total Calcium and Magnesium Carbonate ($CaCO_3$ and $MgCO_3$): 98%
   Total Insolubles: <2%

Piqua Minerals Filler 70:
   Specific Gravity (gm/cc): 2.71
   Median Particle Dia. (microns, Sedigraph): 7
   Wt. per Gallon (lbs, solid): 22.50
   pH of Saturated Solution: 9.40
   Index of Refraction: 1.60
   Dry Brightness (Hunter): 80-82
   Oil Absorption (gm/100 gm): 16-20
   Mohs Hardness: 3
   Bulk Density (lbs/cu ft, loose, approx.): 35
   Free Moisture @ 110C: 0.20 max.
   Percent Passing 325 Mesh Screen: 97
Typical Chemistry:
   Total Calcium and Magnesium Carbonate ($CaCO_3$ and $MgCO_3$): 98%
   Total Insolubles: <2%

Piqua Minerals: PIQUA Minerals Fillers (Continued):

Piqua Minerals Filler 300:
  Specific Gravity (gm/cc): 2.71
  Median Particle Dia. (microns, Sedigraph): 19
  Wt. per Gallon (lbs, solid): 22.50
  pH of Saturated Solution: 9.40
  Index of Refraction: 1.60
  Dry Brightness (Hunter): 78-80
  Oil Absorption (gm/100 gm): 12-16
  Mohs Hardness: 3
  Bulk Density (lbs/cu ft, loose, approx.): 53
  Free Moisture @ 110C: (%): 0.20 max.
  Percent Passing 325 Mesh Screen: 60
Typical Chemistry:
  Total Calcium and Magnesium Carbonate (CaCO3 and MgCO3): 98%
  Total Insolubles: <2%
A fine sized dry ground carbonate filler

Piqua Minerals Filler 600:
  Specific Gravity (gm/cc): 2.71
  Median Particle Dia. (microns, Sedigraph): 53
  Wt. per Gallon (lbs, solid): 22.50
  pH of Saturated Solution: 9.40
  Index of Refraction: 1.60
  Mohs Hardness: 3
  Bulk Density (lbs/cu ft, loose, approx.): 69
  Free Moisture @ 110C (%): 0.20 max
  Percent Passing 325 Mesh Screen: 45
Typical Chemistry:
  Total Calcium and Magnesium Carbonate (CaCO3 and MgCO3): 98%
  Total Insolubles: <2%
A medium sized dry ground carbonate filler.

Potters Industries Inc.: SPHERIGLASS Solid Glass Spheres:
High Performance Solid Glass Polymer Additives

Solid glass spheres provide a unique additive for thermoplastic and thermosetting resin systems. Their multiple benefits, including enhanced processing and reduced manufacturing costs are outlined below.

Glass spheres are smooth, hard and offer excellent chemical resistance and low oil absorption. These and other characteristics enable the spheres to be used in a wide range of applications in the automotive, chemical, electronic, industrial, engineering and photographic industries, where they can substantially reduce reject rates in production.

Multiple Benefits of Spheres:
Process:
* solid smooth shape
* lowest surface to volume ratio
* high loading capacity
* improved lubricity
* low resin mix viscosity
* excellent mold flow
* uniform dispersion
Product:
* low uniform shrinkage
* low warpage
* high flexural modulus
* high abrasion resistance
* high compressive strength
* increased surface hardness
* better stress distribution

Improved Flow Properties:
Spheres lower the viscosity of most resin mix systems, acting as miniature ball bearings to improve flow.
High Resin Displacement:
The precise geometry of Spheriglass spheres allows them to disperse evenly, pack closely, and wet out easily in the compound, permitting very high filler loadings.
Low Shrinkage and Warpage:
High loadings of glass spheres add significantly to the dimensional stability of finished products by reducing shrinkage and improving part flatness.
Better Molded Parts:
Glass spheres produce superior finished product characteristics in many resin systems.
Dimensional Stability:
Better stress distribution is achieved from the use of a spherically shaped particle.

Potters Industries Inc.: SPHERIGLASS Solid Glass Spheres (Continued):

A-Glass:
1922:
    Particle Size Distribution (microns): Mean Value: 203
    Bulk Density (lbs./cu.ft.): Untapped: 91
    Bulk Density (lbs./cu.ft.): Tapped: 98
    Oil Absorption (g oil/100 g spheres): 18
2024:
    Particle Size Distribution (microns): Mean Value: 156
    Bulk Density (lbs./cu.ft.): Untapped: 91
    Bulk Density (lbs./cu.ft.): Tapped: 98
    Oil Absorption (g oil/100 g spheres): 18
2227:
    Particle Size Distribution (microns): Mean Value: 119
    Bulk Density (lbs./cu.ft.): Untapped: 91
    Bulk Density (lbs./cu.ft.): Tapped: 98
    Oil Absorption (g oil/100 g spheres): 18
2429:
    Particle Size Distribution (microns): Mean Value: 85
    Bulk Density (lbs./cu.ft.): Untapped: 91
    Bulk Density (lbs./cu.ft.): Tapped: 98
    Oil Absorption (g oil/100 g spheres): 18
2530:
    Particle Size Distribution (microns): Mean Value: 71
    Bulk Density (lbs./cu.ft.): Untapped: 91
    Bulk Density (lbs./cu.ft.): Tapped: 98
    Oil Absorption (g oil/100 g spheres): 18
2900:
    Particle Size Distribution (microns): Mean Value: 42
    Bulk Density (lbs./cu.ft.): Untapped: 91
    Bulk Density (lbs./cu.ft.): Tapped: 98
    Oil Absorption (g oil/100 g spheres): 18
3000:
    Particle Size Distribution (microns): Mean Value: 35
    Bulk Density (lbs./cu.ft.): Untapped: 80
    Bulk Density (lbs./cu.ft.): Tapped: 99
    Oil Absorption (g oil/100 g spheres): 18

E-Glass:
3000E:
    Particle Size Distribution (microns): Mean Value: 35
    Bulk Density (lbs./cu.ft.): Untapped: 82
    Bulk Density (lbs./cu.ft.): Tapped: 101
    Oil Absorption (g oil/100 g spheres): 17

Typical Values:	A-Glass Soda-Lime	E-Glass Boro-Silicate
Specific Gravity	2.5	2.54
Refractive Index	1.51	1.55
Free Iron Content, % max	0.1	0.1

Shamokin Filler Co., Inc.: CARB-O-FIL:

Typical Analysis:
Chemical Analysis:
   Carbon: 75.0%
   Ash: 18.0%
   Volatile Matter: 7.0%
   Moisture: 1.5%
   Specific Gravity: 1.47+-0.02%

Composition of Volatile Matter:
   Carbon Dioxide: 2.0%
   Oxygen: 2.2%
   Hydrogen: 80.8%
   Nitrogen: 6.6%
   Carbon Monoxide: 6.2%
   Methane: 2.2%

Composition of Ash:
   Silica: 55.00%
   Alumina: 38.12%
   Ferric Oxide: 2.06%
   Titanium Oxide: 1.82%
   Manganous Oxide: 0.03%
   Calcium Oxide: 1.20%
   Magnesium Oxide: 0.60%
   Sodium Oxide + Potassium Oxide: 0.44%
   Phosphorous Pentoxide: 0.13%
   Sulfur Trioxide: 0.60%

Physical Analysis:
   100: 1.0%
   200: 9.0%
   325: 19.0%
   -325: 71.0%

Silbrico Corp.: RYOLEX Filler 23-M:

Chemical Analysis (in elemental form):
    Silicon: 33.8%
    Aluminum: 7.2%
    Potassium: 3.5%
    Sodium: 3.4%
    Calcium: 0.6%
    Iron: 0.6%
    Magnesium: 0.2%
    Trace Elements: 0.2%
    Oxygen (by difference): 47.5%
    Ignition Loss (water): 3.0%

Effective Specific Gravity (g/cm3): 0.8
Percent Free Moisture: <0.5%
Surface pH: 7.0
Color: White
Dry Bulk Density, Lbs./cu.ft.: 6.0
Fusion Point (F): 2,300
Bag Weights: 24 lbs.+-2 lbs.

Particle Size Distribution:

U.S. Sieve	% Weight
+50	Trace
-50 + 100	Max. 10

Unimin Corp.: Premium Mineral Fillers and Extenders:

MINEX Functional Fillers and Extenders-Nepheline Syenite:
   Produced from a unique, silica deficient sodium-potassium-
alumino silicate with a distinctly blue-white hue. This premium
mineral filler offers an exceptionally high dry brightness and
whiteness, which combined with the absence of subtle pink, cream
or buff undertones, will produce the truest representation of
color. Its low tint strength allows for the maximum development
of sharp deeptone colors without any trace of milkiness. Even at
high loadings, Minex functions as a translucent filler to comp-
lement the absorption and light scattering coefficients of the
prime pigments.
   Minex exhibits excellent tint retention over prolonged
exterior exposures without yellowing or deteriorating. Its molecular structure provides excellent chemical and stain resistance
in both exterior and interior architectural coatings, and its
particle geometry provides tight, rigid packing to improve the
integrity of dry film. On surfaces of varying porosity, Minex
fillers produce superior color uniformity and low angle sheen
control. Paints that contain Minex will also remain remarkably
clean. They will not collect and retain dirt, degrade or chalk,
nor hold surface moisture causing the propagation of mildew.
   Minex features an oil absorption value which makes possible
higher bulking levels without compromise to either viscosity
or thixotropy. The Minex particle is easily wetted and dispersed,
and its alkaline pH serves as a buffering and stabilizing agent
for longer shelf life. When product labeling is an issue, Minex
offers formulators the only naturally occurring silicate filler
with no detectable crystalline silica and freedom from OSHA,
NIOSH and NTP silica safety and health labeling. Combined with
proven durability, Minex is an ideal filler in white, pastel
and deeptone trade paints, semi-transparent stains, powder
coatings and OEM finishes.

IMSIL and TAMSIL Microcrystalline Silica Fillers:
   Imsil and Tamsil microcrystalline silica fillers may be the
industry's most versatile fillers. A uniquely formed alpha
quartz with a morphology similar to grape-like clusters, it is
easily wetted and dispersed in both solvent and water-based
systems. Formulators specify Imsil and Tamsil for its excellent
tint retention, durability over prolonged exposure and resistance to dirt, mildew and cracking.
   Interior architectural coatings formulated with Imsil and
Tamsil exhibit strengthened film integrity, increased mar
and stain resistance and more uniform sheen control. Low oil
absorption, regardless of particle size, enables formulators
to increase pigment loading for improved dry hide without
adversely affecting film properties. Conversely, the lower
binder demand of Imsil and Tamsil enables chemists to increase
solids and reduce VOC's without sacrificing the original base
properties.

Unimin Corp.: Premium Mineral Fillers and Extenders (Continued):

IMSIL and TAMSIL Microcrystalline Silica Fillers (Continued):
Exterior architectural coatings will resist acid rain and weathering better than with other fillers because Imsil and Tamsil are completely inert. These microcrystalline silicas reinforce resin resistance to UV rays, help prevent under eave frosting and make possible non-chalking performance. Hardness and chemical purity also make Imsil and Tamsil excellent choices in industrial and marine applications. When used in primers, microcrystalline silica provides 'tooth' for better intercoat adhesion and durability for longer service in abrasive and corrosive environments. In OEM and trade paints, Imsil and Tamsil enhance critical performance properties like water spotting and stain resistance, adhesion, lap sheen, leveling and scrubbability.

SILVERBOND Ground Crystalline Silica:
SilverBond ground crystalline silica is the performance standard by which other crystalline fillers are measured. Completely inert, SilverBond provides excellent tint retention in exterior architectural finishes and helps prevent atmospheric degradation at the surface interface. Low oil absorption allows for high loadings, and its neutral pH makes SilverBond the preferred filler in catalyzed two-component systems. A hard filler, SilverBond imparts excellent abrasion and scratch resistance when incorporated into industrial paint and coatings, grouts, elastomeric sealants and high performence epoxies.

GRANUSIL Mineral Fillers and MINISPHERES Spherical Fillers:
Granusil mineral fillers are the building blocks of emulsion, elastomeric, cemented and modified cementitious systems. Consistently uniform grain shapes and size distributions produce excellent structural and mechanical properties in antiskid coatings. Low levels of soluble alkali and alkaline oxides reduce total conductivity to slow the electrolytic process in corrosive environments.
Minispheres spherical fillers offer increased flow and placement capabilities to yield improved compressive and impact strength. The 2000 Series improves dimensional rigidity without a corresponding loss in flexural or thermal properties. Low surface area and uniform particle distribution add stability to polymer overlays and cast composites. The 4000 Series is designed for high loading and resin displacement in self leveling and troweled polymer overlays and high solids paints. Low oil absorption and improved packing density add volume without weight or shrinkage.

Unimin Corp.: Premium Mineral Fillers and Extenders (Continued):

UNISPAR Micronized Feldspar Fillers:
   Unispar micronized feldspar fillers are processed from a
naturally white, chemically inert feldspar with exceptionally
low soluble salts. The result is a superior, reinforcing mineral
filler which is particularly effective in strengthening and
enhancing a variety of industrial paint systems ranging from
maintenance coatings to traffic paints. An economical workhorse,
Unispar offers formulators many critical performance properties
to help achieve a balance between performance and cost.
   Hard, angular Unispar particles create a rigid packing
network to produce a durable dry film which resists blistering,
blooming or chemical degradation in the harshest industrial
and marine environments. Its high Mohs hardness helps create
tough coatings with excellent abrasion properties to deliver
better stain and burnish resistance then comparable fillers.
In architectural applications, Unispar's combination of size
distribution and chemical purity also offers superior
flatting efficiency and frosting resistance with no cross
reaction.
   Low oil absorption permits the chemist to use Unispar to
increase loading without adversely affecting viscosity. High
loading levels also allow formulators to maximize color strength
and intensity with smaller additions of more expensive coloring
pigments. Unispar fillers will deliver cleaner colors and more
durable performance in commercial, industrial and marine
coatings and linings.

**Unimin Corp.: MINEX Functional Fillers and Extenders: Particle
Size Analysis & Properties:**

2:
 % Finer: 75u: 99.8
 Hegman Value: ASTM D1210-79: N/A
 Median Particle Size: Sedigraph: 14.3
 Specific Surface Area: Fisher Sub Sieve: 0.4
 Brightness: Tappi: 85.9
 Moisture %: ASTM C-566: 0.04
 Oil Absorption: ASTM D-281: 22.5
 pH: AFS 113-87-S: 9.7

3:
 % Finer: 45u: 98.4
 Hegman Value: ASTM D1210-79: N/A
 Median Particle Size: Sedigraph: 10.8
 Specific Surface Area: Fisher Sub Sieve: 0.6
 Brightness: Tappi: 86.8
 Moisture %: ASTM C-56: 0.05
 Oil Absorption: ASTM D-281: 25.0
 pH: AFS 113-87-S: 9.9

4:
 % Finer: 20u: 91.6
 Hegman Value: ASTM D1210-79: 4.6
 Median Particle Size: Sedigraph: 6.8
 Specific Surface Area: Fisher Sub Sieve: 0.7
 Brightness: Tappi: 88.0
 Moisture %: ASTM C-566: 0.07
 Oil Absorption: ASTM D-281: 26.6
 pH: AFS 113-87-S: 10.1

7:
 % Finer: 16u: 98.2
 Hegman Value: ASTM D1210-79: 5.8
 Median Particle Size: Sedigraph: 3.5
 Specific Surface Area: Fisher Sub Sieve: 1.2
 Brightness: Tappi: 89.2
 Moisture %: ASTM C-566: 0.11
 Oil Absorption: ASTM D-281: 31.0
 pH: AFS 113-87-S: 10.2

10:
 % Finer: 10u: 81.0
 Hegman Value: ASTM D1210-79: 6.4
 Median Particle Size: Sedigraph: 2.1
 Specific Surface Area: Fisher Sub Sieve: 1.7
 Brightness: Tappi: 89.7
 Moisture %: ASTM C-566: 0.15
 Oil Absorption: ASTM D-281: 34.1
 pH: AFS 113-87-S: 10.3

U.S. Silica Co.: MIN-U-SIL Fine Ground Silicas:

Min-U-Sil 5 Fine Ground Silica:
  Plant: Berkeley Springs, West Virginia
Typical Physical Properties:
  Bulk Density-Compacted (lbs/ft3): 41   pH: 6.2
  Bulk Density-Uncompacted (lbs/ft3): 36   -5 Micron (%): 97
  Hardness (Mohs): 7   +325 Mesh (%): 0.005
  Hegman: 7.5   Reflectance (%): 92
  Median Diameter (Microns): 1.7   Yellowness Index: 4.2
  Oil Absorption (D-1483): 44   Specific Gravity (g/cm3): 2.65
Typical Chemical Analysis, %:
  SiO2 (Silicon Dioxide): 98.3
  Fe2O3 (Iron Oxide): 0.06
  Al2O3 (Aluminum Oxide): 1.1

Min-U-Sil 10 Fine Ground Silica:
  Plant: Berkeley Heights, West Virginia
Typical Physical Properties:
  Bulk Density-Compacted (lbs/ft3): 45   pH: 6.8
  Bulk Density-Uncompacted (lbs/ft3): 42   -10 Micron (%): 97
  Hardness (Mohs): 7   +325 Mesh (%): 0.005
  Hegman: 7.0-7.5   Reflectance (%): 92
  Median Diameter (Microns): 3.4   Yellowness Index: 3.2
  Oil Absorption (D-1483): 33   Specific Gravity (g/cm3): 2.65
Typical Chemical Analysis, %:
  SiO2 (Silicon Dioxide): 98.6
  Fe2O3 (Iron Oxide): 0.03
  Al2O3 (Aluminum Oxide): 1.0

Min-U-Sil 15 Fine Ground Silica:
  Plant: Pacific, Missouri
Typical Physical Properties:
  Bulk Density-Compacted (lbs/ft3): 54   pH: 7.5
  Bulk Density-Uncompacted (lbs/ft3): 41   -15 Micron (%): 98
  Hardness (Mohs): 7   +325 Mesh (%): 0.003
  Hegman: 6.5   Reflectance (%): 92.0
  Median Diameter (Microns): 4.12   Yellowness Index: 2.0
  Oil Absorption (D-1483): 44   Specific Gravity (g/cm3): 2.65
Typical Chemical Analysis, %:
  SiO2 (Silicon Dioxide): 99.4
  Fe2O3 (Iron Oxide): 0.017
  Al2O3 (Aluminum Oxide): 0.350

U.S. Silica Co.: MIN-U-SIL Fine Ground Silicas (Continued):

Min-U-Sil 30 Fine Ground Silica:
   Plant: Pacific, Missouri
Typical Physical Properties:
   Bulk Density-Compacted (lbs/ft3): 54   pH: 7.5
   Bulk Density-Uncompacted (lbs/ft3): 41   -30 Micron (%): 98
   Hardness (Mohs): 7   +325 Mesh (%): 0.230
   Hegman: 5.5   Reflectance (%): 91.0
   Median Diameter (Microns): 6.90   Yellowness Index: 2.7
   Oil Absorption (D-1483): 44   Specific Gravity (g/cm3): 2.65
Typical Chemical Analysis, %:
   SiO2 (Silicon Dioxide): 99.6
   Fe2O3 (Iron Oxide): 0.017
   Al2O3 (Aluminum Oxide): 0.350

Min-U-Sil 40 Fine Ground Silica:
   Plant: Pacific, Missouri
Typical Physical Properties:
   Bulk Density-Compacted (lbs/ft3): 67   pH: 7.5
   Bulk Density-Uncompacted (lbs/ft3): 50   -40 Micron (%): 98
   Hardness (Mohs): 7   +325 Mesh (%): 1.0
   Hegman: 4.5   Reflectance (%): 90.5
   Median Diameter (Microns): 8.71   Yellowness Index: 3.0
   Oil Absorption (D-1483): 44   Specific Gravity (g/cm3): 2.65
Typical Chemical Analysis, %:
   SiO2 (Silicon Dioxide): 99.6
   Fe2O3 (Iron Oxide): 0.014
   Al2O3 (Aluminum Oxide): 0.230

U.S. Silica Co.: SIL-CO-SIL Ground Silicas:

Sil-Co-Sil 45 Ground Silica:
    Plant: Ottawa, Illinois
    USA Std Sieve Size: 140--106 Microns--% Passing: 100.0
    USA Std Sieve Size: 200-- 75 Microns--% Passing: 99.8
    USA Std Sieve Size: 270-- 53 Microns--% Passing: 99.0
    USA Std Sieve Size: 325-- 45 Microns--% Passing: 98.5
Typical Physical Properties:
    Hardness (Mohs): 7.0   Reflectance (%): 80
    Hegman: 3.5   Yellowness Index: 3.7
    pH: 7.0   Specific Gravity (g/cc): 2.65
Typical Chemical Analysis, %:
    SiO2 (Silicon Dioxide): 99.8

Sil-Co-Sil 52 Ground Silica:
    Plant: Berkeley Springs, West Virginia
    USA Std Sieve Size: 100--150 Microns--% Passing: 100.0
    USA Std Sieve Size: 140--106 Microns--% Passing: 100.0
    USA Std Sieve Size: 200-- 75 Microns--% Passing: 99.7
    USA Std Sieve Size: 270-- 53 Microns--% Passing: 98.0
    USA Std Sieve Size: 325-- 45 Microns--% Passing: 95.5
Typical Physical Properties:
    Hardness (Mohs): 7.0   Reflectance (%): 89
    Mineral: Quartz   Yellowness Index: 3.8
    pH: 7   Specific Gravity (g/cm3): 2.65
Typical Chemical Analysis, %:
    SiO2 (Silicon Dioxide): 99.5
    Fe2O3 (Iron Oxide): 0.022
    Al2O3 (Aluminum Oxide): 0.3

Sil-Co-Sil 53 Ground Silica:
    Plant: Mill Creek, Oklahoma
    USA Std Sieve Size: 100--150 Microns--% Passing: 100.0
    USA Std Sieve Size: 140--106 Microns--% Passing: 100.0
    USA Std Sieve Size: 200-- 75 Microns--% Passing: 99.5
    USA Std Sieve Size: 270-- 53 Microns--% Passing: 96.2
    USA Std Sieve Size: 325-- 45 Microns--% Passing: 92.9
Typical Physical Properties:
    Hardness (Mohs): 7.0   Reflectance (%): 90.4
    Melting Point (Degrees F): 3100 Yellowness Index: 3.4
    Mineral: Quartz   Specific Gravity (g/cm3): 2.65
    pH: 7.0
Typical Chemical Analysis, %:
    SiO2 (Silicon Dioxide): 99.7
    Fe2O3 (Iron Oxide): 0.016
    Al2O3 (Aluminum Oxide): 0.145

U.S. Silica Co.: SIL-CO-SIL Ground Silicas (Continued):

Sil-Co-Sil 63 Ground Silica:
    Plant: Berkeley Springs, West Virginia
    USA Std Sieve Size: 100--Millimeters: 0.150--% Passing: 100.0
    USA Std Sieve Size: 140--Millimeters: 0.106--% Passing: 99.9
    USA Std Sieve Size: 200--Millimeters: 0.075--% Passing: 99.5
    USA Std Sieve Size: 270--Millimeters: 0.053--% Passing: 97.0
    USA Std Sieve Size: 325--Millimeters: 0.045--% Passing: 94.0
Typical Physical Properties:
    Hardness (Mohs): 7.0   Reflectance (%): 88.5
    Mineral: Quartz   Yellowness Index: 4.0
    pH: 7.0   Specific Gravity (g/cm3): 2.65
Typical Chemical Analysis, %:
    SiO2 (Silicon Dioxide): 99.5
    Fe2O3 (Iron Oxide): 0.022
    Al2O3 (Aluminum Oxide): 0.3

Sil-Co-Sil 75 Ground Silica:
    Plant: Berkeley Springs, West Virginia
    USA Std Sieve Size: 100--Millimeters: 0.150--% Passing: 100.0
    USA Std Sieve Size: 140--Millimeters: 0.106--% Passing: 99.8
    USA Std Sieve Size: 200--Millimeters: 0.075--% Passing: 98.5
    USA Std Sieve Size: 270--Millimeters: 0.053--% Passing: 93.0
    USA Std Sieve Size: 325--Millimeters: 0.045--% Passing: 88.0
Typical Physical Properties:
    Hardness (Mohs): 7.0   Reflectance (%): 88
    Mineral: Quartz   Yellowness Index: 4.3
    pH: 7.0   Specific Gravity (g/cm3): 2.65
Typical Chemical Analysis, %:
    SiO2 (Silicon Dioxide): 99.5
    Fe2O3 (Iron Oxide): 0.022
    Al2O3 (Aluminum Oxide): 0.3

Sil-Co-Sil 90 Ground Silica:
    Plant: Berkeley Springs, West Virginia
    USA Std Sieve Size: 70: Millimeters: 0.212--% Passing: 100.0
    USA Std Sieve Size: 100: Millimeters: 0.150--% Passing: 99.8
    USA Std Sieve Size: 140: Millimeters: 0.106--% Passing: 98.7
    USA Std Sieve Size: 200: Millimeters: 0.075--% Passing: 94.8
    USA Std Sieve Size: 270: Millimeters: 0.053--% Passing: 86.0
    USA Std Sieve Size: 325: Millimeters: 0.045--% Passing: 80.0
Typical Physical Properties:
    Hardness (Mohs): 7.0   Reflectance (%): 87.5
    Mineral: Quartz   Yellowness Index: 4.3
    pH: 7.0   Specific Gravity (g/cm3): 2.65
Typical Chemical Analysis, %:
    SiO2 (Silicon Dioxide): 99.5
    Fe2O3 (Iron Oxide): 0.022
    Al2O3 (Aluminum Oxide): 0.3

U.S. Silica Co.: SIL-CO-SIL Ground Silicas (Continued):

Sil-Co-Sil 106 Ground Silica:
Plant: Mill Creek, Oklahoma
USA Std Sieve Size: 70--Millimeters: 0.212--% Passing: 100.0
USA Std Sieve Size: 100--Millimeters: 0.150--% Passing: 99.9
USA Std Sieve Size: 140--Millimeters: 0.106--% Passing: 99.0
USA Std Sieve Size: 200--Millimeters: 0.075--% Passing: 95.1
USA Std Sieve Size: 270--Millimeters: 0.053--% Passing: 84.4
USA Std Sieve Size: 325--Millimeters: 0.045--% Passing: 77.7
Typical Physical Properties:
Hardness (Mohs): 7.0   Reflectance (%): 89.4
Melting Point (Degrees F): 3100 Yellowness Index: 3.63
Mineral: Quartz   Specific Gravity (g/cm3): 2.65
pH: 7.0
Typical Chemical Analysis:
SiO2 (Silicon Dioxide): 99.7
Fe2O3 (Iron Oxide): 0.016
Al2O3 (Aluminum Oxide): 0.135

Section IX

Fire and Flame Retardants/ Smoke Suppressants

Aceto Corp.: Trichloro Ethyl Phosphate: Tris (beta-chlorethyl) Phosphate: (CLCH2CH2O)3PO:

Trichloro ethyl phosphate is not only a most effective fire retardant plasticizer, but can also be utilized in transparent films. This very unique combination of properties has resulted in the use of this plasticizer in a great variety of plastics and a great number of product applications.

Trichloro ethyl phosphate can be used with PVC, PVA, polyesters, polyurethanes, phenolics, epoxies, cellulose acetate, ethyl cellulose and others. The use level depends on the application and has been reported to range between 3% and 20%.

Typical Physical Properties:
Molecular Weight: 285.51
Boiling Point @ 760 mm Hg: 220C (with decomposition)
Boiling Point @ 25 mm Hg: 214C
Specific Gravity (20/4C): 1.4256
Refractive Index (nD20): 1.4731
Viscosity @ 20C: 42.9 cps
Flash Point: 230C
Solubility in Water: 0.05% w/w
Solubility in Water in TCEP: 5.5% w/w

Specifications:
Color: 50 APHA
Water Content: 0.10% w/w maximum
Assay (KOH/g): 0.05 maximum
Chloride Content as Cl (after heating 5 hrs.
@ 100C): 0.01% w/w maximum
Volatility (loss after heating 5 hrs. @ 100C): 0.5% w/w max
Acidity (after heating 5 hrs. @ 100C)
(gms. NaOH/100 gms): 0.05 maximum
pH of Aqueous Extract (10 vols. water to
1 vol. TCEP): no less than 5
Specific gravity @ 20/20C: 1.415-1.435
Distillation Range at 200 mm Hg: 210-220C
Refractive Index (nd20): 1.470-1.480

Suggested Uses:
Aceto's Trichloro Ethyl Phosphate is used as a flame retardant plasticizer in such resins as:
Cellulose Acetate
Cellulose Acetate Butyrate
Ethyl Cellulose
Modified Urea-Formaldehyde
Nitrocellulose
Phenol-Formaldehyde
Polyesters
Polyurethanes
Shellac

Its use in acrylic polymers and copolymers, butadiene styrene latices, epoxy resins and polystyrene is also recommended.

Albemarle Corp.: SAYTEX Flame Retardants:

Saytex RB-100:
Chemical Name: Tetrabromo bisPhenol-A
Form: White powder
% Br: 59

Saytex CP-2000:
Chemical Name: Tetrabromo bisPhenol-A
Form: White powder
% Br: 59

Saytex 102E:
Chemical Name: Decabromo-diphenyl oxide
Form: White powder
% Br: 83

Saytex 8010:
Chemical Name: Proprietary
Form: White powder
% Br: 82

Saytex BT-93:
Chemical Name: Ethylene bis-tetrabromo phthalimide
Form: Yellow powder
% Br: 67

Saytex BT-93W:
Chemical Name: Ethylene bis-tetrabromo phthalimide
Form: White powder
% Br: 67

Saytex 120:
Chemical Name: Tetradeca-bromodiphenoxy benzene
Form: White powder
% Br: 82

Saytex HP-7010 P/G:
Chemical Name: Polymeric FR
Form: White powder
% Br: 69

Saytex HP-800:
Chemical Name: Tetrabromo-bisphenol-A-bis (2,3-dibromopropyl
                ether
Form: White powder
% Br: 68

Saytex HBCD:
Chemical Name: Hexabromo-cyclo-dodecane
Form: White powder
% Br: 75

Albemarle Corp.: SAYTEX Flame Retardants (Continued):

Saytex HP-900:
   Chemical Name: Hexabromo-cyclo-dodecane
   Form: White powder
   % Br: 75

Saytex BC-48:
   Chemical Name: Tetrabromo cyclooctane
   Form: White powder
   % Br: 75

Saytex BCL-462:
   Chemical Name: Dibromoethyl-dibromo-cyclohexane
   Form: White powder
   % Br: 75

Saytex RB-49:
   Chemical Name: Tetrabromo phthalic anhydride
   Form: White powder
   % Br: 69

Saytex RB-79:
   Chemical Name: Mixture
   Form: Amber liquid
   % Br: 45

Saytex RB-7980:
   Chemical Name: RB-79/Phosphate ester blend
   Form: Amber liquid
   % Br: 36

Albright & Wilson Americas Inc.: ANTIBLAZE NK Flame Retardant:

Antiblaze NK Flame Retardant represents a significant advance in polyolefin flame retardant additives. Fire Retardant thermoplastics with low density, high elongation and high impact strength with a low tint strength additive are now available. Antiblaze NK is a major improvement over existing olefin flame retardants due to its high efficiency without the need to add other 'synergists'. Antiblaze NK is a neutral compound in aqueous systems, and is therefore compatible with co-additives used in curable systems such as coatings, urethanes and epoxies.

Suggested Uses:
Thermosets & thermoplastics
Olefin Polymers and Copolymers, Polyester Epoxies
EVA, TPE & TPU used in Extrusion, Molding & Film Applications

Performance Benefits:
* Very high phosphoric acid content
* High efficiency - UL94 VO in polypropylene at 35% loadings
* Small particle size for maximum efficiency and retention of physical properties in finished goods
* Excellent thermal stability
* Neutral pH in aqueous systems

Typical Chemical/Physical Properties:
Appearance: White, free flowing powder
Density: 1.22 g/cc
Bulk Density: 0.63 g/cc
Phosphorus Content: 19.6%
H3PO4 Content: 62 Wt%
Melting Point: Decomposes above 250C
Particle Size: Nominal 10 microns
Thermal Stability Data: 250C: 2% Weight Loss
  (TGA)   300C: 15% Weight Loss
                          400C: 30% Weight Loss

Albright & Wilson Americas Inc.: ANTIBLAZE NP Flame Retardant:

Antiblaze NP Flame Retardant represents a significant advance in polyolefin flame retardants. Fire retardant polyolefins with low density, high elongation and high impact strength are now available with an additive which does not impair melt flow properties. Antiblaze NP is recommended for use in polypropylene homopolymer and copolymers, high and low density polyethylene and ethylene-vinyl acetate compolymers. Antiblaze NP is a major improvement over other flame retardants due to its 'high effic-iency' without the need for additive 'synergists'.

Suggested Uses:
Thermoplastics & Thermosets
Olefin Polymers and Copolymers, Polyesters, Epoxies
TPE & TPU used in Extrusion, Molding & Sheet

Performance Benefits:
* Very high phosphoric acid content
* High efficiency - UL94 VO at 1/16" in polypropylene at 30% loadings
* Small particle size for maximum efficiency and retention of physical properties in finished goods
* Provides flame retardant plastics without water sensitivity
* Excellent thermal stability
* Neutral pH in aqueous systems

Typical Chemical/Physical Properties:
Appearance: White, free flowing powder
Density: 1.20 g/cc
Bulk Density: 0.63 g/cc
Phosphorus Content: 15.7%
H_3PO_4 Content: 50 Wt%
Melting Point: Decomposes above 250C
Particle Size: 50% median 10 microns
Thermal Stability Data: 250C: 1% Weight Loss
  (TGA)   300C: 9% Weight Loss
                                 400C: 53% Weight Loss

Albright & Wilson Americas Inc.: ANTIBLAZE 1045 Flame Retardant:

Antiblaze 1045 Flame Retardant is a unique phosphorus based
product intended for the plastics industry. Both thermoplastic
and thermoset applications with demanding performance requirements can benefit from the use of this additive product. The
balance of properties which characterize Antiblaze 1045 Flame
Retardant are unique.

Performance Benefits:
 * Highly efficient due to high phosphorus content - 21%
 * A high molecular weight liquid product without particulate
   solids
 * High thermal stability and low volatility
 * Compatible with many polymers

Typical Properties:
Active Ingredient: 100%
Appearance: Glass type liquid
Phosphorus Content: 20.8%
Acid Number, mg KOH/gm: <2
Density, 25C: 1.26 gm/cc
Flash Point, Closed Cup: 249C
Viscosity: 25C: 1,500,000 cPs
             60C: 6,700 cPs
            100C: 2,000 cPs
Solubility: Water: Miscible
            Acetone: Miscible
            Ethanol: Miscible
            Pentane: <5
            Toluene: >10
Volatility: 200C: <1
 TGA Loss: 250C: 2
            300C: 7
            350C: 13

Applications Data:
Recommended applications of Antiblaze 1045 Flame Retardant
include epoxies, unfilled thermoplastic polyesters, polycarbonates and selected polyamides. Reinforced epoxy composites
with excellent physical properties can be achieved with 8-15%
Antiblaze 1045 Flame Retardant, depending upon resins and curing
agents. Flame retardant polyethylene terephthalate (PET) can be
produced with excellent physical properties due to the high
efficiency of Antiblaze 1045. Addition levels of 3-5 weight
percent provide UL 94 VO ratings at 1/16". The physical properties of flame retardant PET are only slightly reduced when
compared to PET without flame retardant. There is 12-18% reduction in break strength and elongation. These reductions can be
largely overcome by using better handling techniques or a higher
molecular weight PET.

Alcan Chemicals: Alumina Trihydrate-SF Grades:

The Alcan Superfine "SF" grades of alumina trihydrate are directly precipitated, under tightly controlled conditions, to yield a range of products with low particle size to surface area ratios. These products provide the user with a low toxicity way to retard flame and suppress smoke in a variety of polymeric systems.

Alcan SF grades of alumina trihydrate are white crystalline powders with the chemical formula $Al(OH)_3$, sometimes expressed as $Al_2O_3-3H_2O$.

Alcan SF grades absorb energy and release water vapor on heating, imparting flame retardant and smoke suppressant properties to plastic and rubber products.

Electrical Grades:
Alcan SF 'E' grades have identical physical properties to standard grades, but they have a much lower ionic impurity content. This permits improved performance in electrical applications. In particular the 'E' grades are ideal as fillers in cable compounds where low electrolyte levels are critical.

Surface Treated Grades:
To achieve higher filler loadings, reduced compound viscosity, or improved physical properties, Alcan SF and Alcan SF'E' products can be supplied surface treated. The treatments that are available include silanes, fatty acids and plasticizers.

Modified Flow and Bulk Density Grades:
For automated powder handling systems and to achieve reduced mix cycle times, Alcan SF grades can be supplied with enhanced flow characteristics and in a number of bulk densities.

Benefits:
  *Halogen free
  *Delayed ignition and retarded rate of burning
  *Reduced smoke emission
  *Good electrical performance
  *Improved compounding rheology
  *Easy replacement of other fillers
  *Low toxicity
  *High degree of whiteness
  *Can be supplied surface treated

Typical Applications:
  *Cable compounds
  *Conveyor belting and flooring compounds
  *Thermoplastic moldings, extrusions and sheet
  *PVC and rubber technical goods
  *Fabric coatings
  *Polyester and acrylic pultrusion
  *Thermoset molding compounds
  *Paper filling and coating
  *Printing inks and paint

Alcan Chemicals: Alumina Trihydrate-SF Grades (Continued):

SF2:
    Surface Area-Minimum-m2/g: 2.0   Maximum-m2/g: 3.5
    Median Particle Size: um: 1.6
    Oil Absorption Value: g/100g: 30
SF4:
    Surface Area-Minimum-m2/g: 3.0   Maximum-m2/g: 6.0
    Median Particle Size: um: 1.2
    Oil Absorption Value: g/100g: 35
SF7:
    Surface Area-Minimum-m2/g: 6.0   Maximum-m2/g: 8.0
    Median Particle Size: um: 0.8
    Oil Absorption Value: g/100g: 42
SF9:
    Surface Area-Minimum-m2/g: 8.0   Maximum-m2/g: 10.0
    Median Particle Size: um: 0.6
    Oil Absorption Value: g/100g: 45
SF11:
    Surface Area-Minimum-m2/g: 10.0   Maximum-m2/g: 12.0
    Median Particle Size: um: 0.5
    Oil Absorption Value: g/100g: 46

SF2E:
    Surface Area-Minimum-m2/g: 2.0   Maximum-m2/g: 3.5
    Median Particle Size: um: 1.6
    Oil Absorption Value: g/100g: 30
SF4E:
    Surface Area-Minimum-m2/g: 3.0   Maximum-m2/g: 6.0
    Median Particle Size: um: 1.2
    Oil Absorption Value: g/100g: 35
SF7E:
    Surface Area-Minimum-m2/g: 6.0   Maximum-m2/g: 8.0
    Median Particle Size: um: 0.8
    Oil Absorption Value: g/100g: 42
SF9E:
    Surface Area-Minimum-m2/g: 8.0   Maximum-m2/g: 10.0
    Median Particle Size: um: 0.6
    Oil Absorption Value: g/100g: 45
SF11E:
    Surface Area-Minimum-m2/g: 10.0   Maximum-m2/g: 12.0
    Median Particle Size: um: 0.5
    Oil Absorption Value: g/100g: 46

Chemical Analysis (%) (All Grades):
    Al2O3: 64.9   CaO: 0.01
    H2O (combined): 34.9   SiO2: 0.015
                            Fe2O3: 0.004

Physical Properties (All Grades):
    Specific Gravity: 2.42
    Refractive Index: 1.57
    Mohs' Hardness: 2.5-3.5

Alcan Chemicals: Alumina Trihydrate-SF Super Density Grades:

The Alcan Superfine "SF" grades of alumina trihydrate are directly precipitated, under tightly controlled conditions, to yield a range of products with low particle size to surface area ratios. These products provide the user with a low toxicity way to retard flame and suppress smoke in a variety of polymeric systems.

Alcan SF grades of alumina trihydrate are white crystalline powders with the chemical formula $Al(OH)_3$, sometimes expressed as $Al_2O_3-3H_2O$.

Electrical Grades:
Alcan SF'E' grades have identical physical properties to standard grades, but they have a much lower ionic impurity content. This permits improved performance in electrical applications. In particular the 'E' grades are ideal as fillers in cable compounds where low electrolyte levels are critical.

Super Density Grades:
For automated powder handling systems and to achieve reduced mix cycle times, Alcan SF'E' grades can be supplied at higher densities to achieve enhanced flow characteristics. Dispersion, top-cut and electrical characteristics are maintained.

Benefits:
  *Halogen free
  *Delayed ingnition and retarded rate of burning
  *Reduced smoke emission
  *Good electrical performance
  *Improved compounding rheology
  *Easy replacement of other fillers
  *Low toxicity
  *High degree of whiteness
  *Free flowing

SF4ESD:
  Surface Area-Minimum-m2/g: 3.0   Maximum-m2/g: 6.0
  Median Particle Size: um: 1.2
  Bulk Density, Untamped-g/cm3: 0.45
                Tamped-g/cm3: 0.65
  Oil Absorption Value: g/100g: 35
SF7ESD:
  Surface Area-Minimum-m2/g: 6.0   Maximum-m2/g: 8.0
  Median Particle Size: um: 0.8
  Bulk Density, Untamped-g/cm3: 0.4
                Tamped-g/cm3: 0.6
  Oil Absorption Value: g/100g: 42

Alcan Chemicals: FLAMTARD Flame Retardants:

Flamtard is a low smoke, flame retardant additive that can be used in a wide variety of polymers and applications.

Grades:
Flamtard H (Zinc hydroxystannate-ZHS)
Flamtard S (Zinc stannate-ZS)
Flamtard HB1 (ZHS/Zinc borate blend)

Mechanism:
The Flamtard grades function as both vapor phase and condensed phase flame retardants and smoke suppressants. This dual phase action acts synergistically in halogenated plastic and rubber formulations. For non-halogenated polymers, the addition of Flamtard with a halogen source will improve fire performance and reduce smoke emission.

Benefits:
Excellent flame retardant
Reduces smoke emission
Decreases carbon monoxide evolution
Promotes char formation
Lower in toxicity and tinting strength

Polymers and Applications:
Rigid and flexible PVC
Polyolefins, polyamides and polyesters
Plenum cable jacket compounds
Paints and coatings
Automotive components

Recommended addition levels would depend on flame and smoke test requirements and on the other components within a specific formulation. As a guide, the substitution of an existing flame retardant, such as antimony trioxide, with the appropriate Flamtard grade is normally very effective for halogenated polymer systems.

Flamtard S:
Physical form: White powder
Median particle size (um): 2.5
Specific Gravity (g/cm3): 3.9
Decomposition temperature (C): >400
Flamtard H:
Physical form: White powder
Median particle size (um): 2.5
Specific Gravity (g/cm3): 3.3
Decomposition temperature (C): 200
Flamtard HB1:
Physical form: White powder
Median particle size (um): 1.8
Specific Gravity (g/cm3): 3.0
Decomposition temperature (C): 200

Aluchem Inc.: Al(OH)3 Hydrated Aluminas:

General Characteristics:
    Chemical Formula: Al2O3-3H2O or Al(OH)3
    Specific Gravity: 2.42
    Decomposition Temperature: 220C
    Refractive Index: 1.57
    Mohs' Hardness: 2.5-3.5
    pH (10% Slurry): 8-10
    Appearance: Crystalline Powder

Typical Chemical Compositions:	White Hydrates	Bayer Hydrates
Color	White	Off-White
Color Reflectance	93-98	80-95
Al2O3, %	65.0	65.0
L.O.I. (110-1100C), %	34.5	34.5
SiO2, %	0.01	0.01
Fe2O3, %	0.01	0.01
Na2O (Total), %	0.15	0.30
Na2O (Soluble), %	0.005	0.007

Grade Nomenclature:
     30 Series: Designates unground/screened products
    400 Series: Designates roller milled products; the lower the
               number the finer the grind
    700 Series: Designates air-attrition milled products, the lower
               the number the finer the grind

White Hydrated Aluminas:
    Designated by the number "1" or the letter "W" appearing in
    the last position of the grade number/letter sequence.
Bayer Off-White Hydrated Aluminas:
    Designated by a "zero" or any number other than "1" in the
    last position of the grade number sequence.

Typical Physical Properties:

Unground:	AC-31	AC-30
Color:	White	Off-White
Bulk Density, Loose, lbs/ft3	72	68
Bulk Density, Packed, lbs/ft3	83	80
Free Moisture (110C), %	0.05	0.05
Oil Absorption	----	----
Surface Area, m2/gm	0.2-0.3	0.1-0.2
Screen Analysis:		
% on 100 Mesh	0-7	2-15
% on 200 Mesh	40-80	50-97
% on 325 Mesh	80-100	85-100
% through 325 Mesh	0-20	0-15

Aluchem, Inc.: Al(OH)3 Hydrated Aluminas (Continued):

Typical Physical Properties:

Coarse & Intermediate Grinds:	AC-470	AC-450	AC-440
Color	Off-White	Off-White	Off-White
Bulk Density, Loose, lbs/ft3	67	63	62
Bulk Density, Packed, lbs/ft3	79	79	78
Free Moisture (110C), %	0.2	0.2	0.2
Oil Absorption	23	23	23
Surface Area, m2/gm	0.75-1.5	0.75-1.5	1.25-1.75
Screen Analysis:			
% on 100 Mesh	0	0	0
% on 200 Mesh	9-13	6-10	2-4
% on 325 Mesh	30-35	20-25	15-20
% through 325 Mesh	65-70	75-80	80-85
Median Particle Size, Microns	-----	16-19	14-17

Fine Grinds:	AC-420	AC-410	AC-400
Color	Off-White	Off-White	Off-White
Bulk Density, Loose, lbs/ft3	56	50	41
Bulk Density, Packed, lbs/ft3	78	74	69
Free Moisture (110C), %	0.2	0.3	0.3
Oil Absorption	24	26	28
Surface Area, m2/gm	1.5-2.0	2-3	2-3
Screen Analysis:			
% on 100 Mesh	0	0	0
% on 200 Mesh	<1	0	0
% on 325 Mesh	5-10	2-5	<0.5
% through 325 Mesh	90-95	95-98	99.5+
Median Particle Size, Microns	11-14	9-12	7-10

Superfine Grinds:	AC-720	AC-714	AC-712
Color	Off-White	Off-White	Off-White
Bulk Density, Loose, lbs/ft3	28	25	22
Bulk Density, Packed, lbs/ft3	50	47	44
Free Moisture (110C), %	0.3	0.4	0.5
Oil Absorption	31	36	38
Surface Area, m2/gm	3.5-4.0	4.0-4.5	5.5-6.0
Screen Analysis:			
% on 100 Mesh	0	0	0
% on 200 Mesh	0	0	0
% on 325 Mesh	<0.01	<0.01	<0.01
% through 325 Mesh	99.9	99.9	99.9
Median Particle Size, Microns	3.5-4.0	3.0-3.5	2.5-3.0

Aluchem, Inc.: Al(OH)3 Hydrated Aluminas (Continued):

Typical Physical Properties:

Other Ground Products:	AC-480	AC-460	AC-430
% on 100 Mesh	0-2	0	0
% on 200 Mesh	13-18	6-10	1-2
% on 325 Mesh	35-40	25-30	10-15
% through 325 Mesh	60-65	70-75	85-90
Median Particle Size, Microns	---	-----	13-25

	AC-421	AC-401	AC-400F
% on 100 Mesh	0	0	0
% on 200 Mesh	<1	0	0
% on 325 Mesh	7-12	<0.5	<0.5
% through 325 Mesh	88-93	99.5+	99.5+
Median Particle Size, Microns	13-16	8-10	6.5-8.5

	AC-740	AC-722	AC-714W
% on 100 Mesh	0	0	0
% on 200 Mesh	0	0	0
% on 325 Mesh	<0.5	<0.01	<0.01
% through 325 Mesh	99.5+	99.9	99.9
Median Particle Size, Microns	6-7	3.5-4.5	3.0-3.5

Alusuisse Aluminum USA: MARTINAL and MAGNIFIN Flame Retardant Fillers:

Fine Precipitated Martinal Grades:
OL-111/LE:
    Median Particle Size um: 0.7-1.2
    Specific Surface m2/g: 10-12
    Oil Absorption cm3/100g: 45-60

OL/Q-111:
    Median Particle Size um: 0.7-1.2
    Specific Surface m2/g: 10-12
    Oil Absorption cm3/100g: 52-64

OL-107:
    Median Particle Size um: 0.9-1.5
    Specific Surface m2/g: 6-8
    Oil Absorption cm3/100g: 37-47

OL-107/LE:
    Median Particle Size um: 0.9-1.5
    Specific Surface m2/g: 6-8
    Oil Absorption cm3/100g: 40-55

OL/Q-107:
    Median Particle Size um: 0.9-1.5
    Specific Surface m2/g: 6-8
    Oil Absorption cm3/100g: 40-55

OL-107/C:
    Median Particle Size um: 0.9-1.5
    Specific Surface m2/g: 6-8
    Oil Absorption cm3/100g: 30-39

OL-104:
    Median Particle Size um: 1.3-2.3
    Specific Surface m2/g: 3-5
    Oil Absorption cm3/100g: 27-35

OL-104/LE:
    Median Particle Size um: 1.3-2.3
    Specific Surface m2/g: 3-5
    Oil Absorption cm3/100g: 30-40

OL-104/C:
    Median Particle Size um: 1.3-2.3
    Specific Surface m2/g: 3-5
    Oil Absorption cm3/100g: 24-31

OL-104/S:
    Median Particle Size um: 1.3-2.3
    Specific Surface m2/g: 3-5
    Oil Absorption cm3/100g: 27-35

Alusuisse Aluminum USA: MARTINAL and MAGNIFIN Flame Retardant Fillers (Continued):

Ground Martinal and Low-Viscosity-Grades:
ON-4608:
    Median Particle Size um: 7-10
    Oil Absorption cm3/100g: 24-30
    Bulk Density kg/m3: 620

ON-310:
    Median Particle Size um: 9-12
    Oil Absorption cm3/100g: 23-29
    Bulk Density kg/m3: 640

ON-313:
    Median Particle Size um: 11-15
    Oil Absorption cm3/100g: 23-29
    Bulk Density kg/m3: 720

ON-320:
    Median Particle Size um: 15-25
    Oil Absorption cm3/100g: 20-25
    Bulk Density kg/m3: 900

ON:
    Median Particle Size um: 45-65
    Bulk Density kg/m3: 1100

ON-904:
    Median Particle Size um: 4
    Oil Absorption cm3/100g: 21
    Bulk Density kg/m3: 530

ON-906:
    Median Particle Size um: 6
    Oil Absorption cm3/100g: 18
    Bulk Density kg/m3: 550

ON-908:
    Median Particle Size um: 8
    Oil Absorption cm3/100g: 17
    Bulk Density kg/m3: 600

ON-921:
    Median Particle Size um: 21
    Oil Absorption cm3/100g: 17
    Bulk Density kg/m3: 650

ON-935:
    Median Particle Size um: 35
    Oil Absorption cm3/100g: 16
    Bulk Density kg/m3: 800

Alusuisse Aluminum USA: MARTINAL and MAGNIFIN Flame Retardant Fillers (Continued):

Magnifin Grades:
H5:
    Median Particle Size um: 1.25-1.65
    Specific Surface m2/g: 4.0-6.0
    Bulk Density g/l: >300
    Whiteness %: >96.0
    Electrical Conductivity uS/cm: <350
    Moisture %: <0.3

H7:
    Median Particle Size um: 0.80-1.10
    Specific Surface m2/g: 7.0-9.0
    Bulk Density g/l: >300
    Whiteness %: >96.0
    Electrical Conductivity uS/cm: <350
    Moisture %: <0.3

H10:
    Median Particle Size um: 0.65-0.95
    Specific Surface m2/g: 9.0-11.0
    Bulk Density g/l: >350
    Whiteness %: >96.0
    Electrical Conductivity uS/cm: <350
    Moisture %: <0.3

Amspec Chemical Corp.: Antimony Oxide:

Antimony Oxide is a white, odorless, fine powder. It is usually produced by roasting ores such as stibnite (antimony sulfide) or by oxidizing antimony metal. The principal ore deposits in the world are found in China, South Africa, Bolivia, Mexico and CIS (formerly USSR). China has been the dominant supplier of the world's antimony raw materials in the 1980's and early 1990's. However, after the collapse of Soviet Union, CIS has begun to play a more important role in the world antimony market.
Amspec offers the following grades of Antimony Oxide:

KR Grade:
A high quality grade with excellent whiteness, high opacity and high tinting strength.

KR Grade-Superfine:
A high quality, small particle size grade, suitable for narrow fiber or thin film applications.

LTS Grade:
A special low tinting strength grade for application where color matching is a major concern. The low tinting strength provides a minimum amount of whiting influence, thus permitting use of smaller amounts of pigments or dyes.

AMSTAR Grade:
A utility grade for those applications where antimony oxide color and particle size distribution are not of concern, such as the roofing industry, conveyor belts, and certain textile applications such as theater blackout and curtains or room darkening shades and drapes.

Dampened Oxide:
Custom made for those customers who require it to reduce antimony oxide dust in their plants. The type of oil or plasticizer used is based on each customer's specification and is normally added at 2-4%. Currently antimony oxide is dampened with mineral oil, DIDP, ethylene glycol, chlorinated paraffin...etc.

AMSPEC Select:
This is a high purity grade of antimony oxide for use as a catalyst for making PET or as a chemical intermediate where low levels of lead, arsenic and other impurities are required.

Liquid Antimony Oxide:
This patented new product is to be used in place of straight antimony oxide powder or concentrates. This new product, having antimony oxide loadings of 85% or greater, provides dustless operation, improves the physical characteristics of plastics and reduces waste.

Amspec Chemical Corp.: Antimony Oxide (Continued):

Flame Retarding with Antimony Oxide

All of the following should be considered merely as starting points or guidelines for flame retarding the various materials covered. Obviously, various specific formulations would be necessary to meet the particular end requirements and flame tests associated with each individual application. Considerations in selection of flame retardant compounds would include the following:

1. What tests are to be met?
2. Economics of total system.
3. Effect on physical, electrical, thermal and other properties.
4. Effect on processing conditions.
5. Related governmental regulations.

Plastics:

Polyvinylchloride (PVC):
Antimony oxide is widely used as a very effective flame retardant in plasticized PVC. The usual level of KR grade Sb2O3 used is from 2 to 10 parts per hundred parts of the PVC. The PVC contains more chlorine than needed for the reaction to convert all of the antimony to SbCl3. LTS Grade Sb2O3 is being used with applications, such as film manufacturing, that require low tinting strength from antimony oxide.

Polypropylene (PP):
KR Grade antimony oxide in combination with halogenated compounds is an effective flame retardant for Polypropylene. Typical formulation could be 4% brominated organic and 2% KR Grade SB2O3 for UL-94 V2 rating or 22% brominated organic and 4% KR Grade Sb2O3 and 14% clay for UL-94 V-O rating.

Polyethylene (PE):
Polyethylene is generally easier to flame retard than polypropylene. Effective flame retardancy can be obtained by using KR Grade Sb2O3 and one or more of the many different chlorinated or brominated compounds. Flame retarded polyethylene can be obtained with a 10-40% loading of a mixture of halogenated organic compounds and KR Grade Sb2O3 which can pass many flame tests.

Acrylonitrile-Butadiene-Styrene (ABS):
Antimony oxide in combination with a halogenated organic is an effective flame retardant for ABS. Using aromatic bromine compounds with KR Grade SB2O3 at a 4:1 organo-Br/Sb2O3 ratio and 21% loading yielded a UL-94 V-O result.

Amspec Chemical Corp.: Antimony Oxide (Continued):

Plastics (Continued):
High Impact Polystyrene (HIPS):
   High impact polystyrene can pass the UL-94 V-O rating by adding 12% aromatic bromine with 4% KR Grade Sb2O3.

Polyester:
   Reinforced chlorinated polyesters can meet ASTM E-84 with addition of 5% KR Grade Sb2O3. Nonchlorinated polyester can be flame retarded to pass ASTM D-635 with additions of 10-12 PHR KR Grade Sb2O3 and 20-24 PHR chlorinated hydrocarbon.

Polyurethanes (PU):
   Both flexible and rigid foam can be flame-retarded to pass ASTM D-1692 with the addition of 5-15% KR Grade Sb2O3 and 5-15% halogenated compounds.

Engineering Resins and Polymer Alloys:
   KR Grade Sb2O3 is often being used in conjunction with other flame retardants in the engineering resins and polymer alloys to achieve desired results. Please call Amspec's technical services for recommendations.

Textiles

Paper

Paint

Rubber

D.J. Enterprises, Inc.: FERRO-CHAR B-44M:

Antimony oxide in combination with halogenated additives has been used in the plastic industry for many years to flame retard plastics. By blending synergists, such as Ferro-Char, the total flame retardant system loading can be effective at lower levels.

The result: cost effectiveness, without reducing performance level.

Load levels will vary between 3-12%, depending on required result. Only your laboratory testing procedure will determine your proper addition rate.

Ferro-Char can also be used with glass-reinforced, hollow or solid sphere, MgO, mica, and talc compounds.

Application ranges are broad, thermoplastics, PVC, urethane and other polymeric materials including adhesives, epoxies, and rubber compounds.

Typical Analysis:
```
Fe:          64%
Al2O3:        4%
TiO2:         6%
SiO2:         2%
MgO:          1.00%
V2O5:         1.00%
CaO, Cr2O3: Trace
MnO, P2O5:      "

pH: 6.8-7.2
S.G.: 4.85
Melt P.: 2,400F
Size: 44-75 Mic.
```

Dover Chemical Corp.: DOVERGUARD Brominated and Bromo-Chlorinated Flame Retardants:

Bromine is sometimes the best choice for flame retardant applications because of its high level of effectiveness. Dover Chemical has developed several brominated and bromo-chlorinated aliphatic liquid flame retardants.

The brominated liquid additives are low cost, highly efficient flame retardants and plasticizers with good color. They should be considered as flame retardants for PVC, polyurethane foams, textiles, carpet backing, adhesives, air filters and EPDM rubber.

Doverguard 8207-A:
    Color G: 1
    Poise/25C: 22
    % Br: 30
    % Cl: 29
    SG/50C: 1.42

Doverguard 8307-A:
    Color G: 1
    Poise/25C: 1
    % Br: 24
    % Cl: 23
    % P: 5
    SG/50C: 1.37

Doverguard 8426:
    Color G: 1
    Poise/25C: 0.15
    % Br: 42
    SG/50C: 1.16

Doverguard 9122:
    Color G: 1
    Poise/25C: 3
    % Br: 16
    % Cl: 16
    % P: 4
    SG/50C: 1.35

Doverguard 8208-A:
    Color G: 2
    Poise/25C: 15
    % Br: 26
    % Cl: 32
    % P: 1.2
    SG/50C: 1.38

Doverguard 8410:
    Color G: 3
    Poise/25C: 0.5
    % Br: 57
    SG/50C: 1.52

Doverguard 9119:
    Color G: 1
    Poise/25C: 65
    % Br: 33
    % Cl: 33
    % P: 0
    SG/50C: 1.58

**Dover Chemical Corp.: DOVERSPERSE Aqueous Chlorinated Flame
   Retardants:**

DOVERSPERSE A-1 and DOVERSPERSE 3 typically contain 45%
available chlorine for maximum flame retardant efficiency. They
find application in both cationic and anionic emulsion systems
because of their non-ionic bases.
   In addition to contributing flame retardancy, they improve
adhesion, impart chemical and water resistance, and allow the
user to formulate aqueous rather than solvent systems.
   Doversperse A-1 is recommended if increased hardness is required. Doversperse 3 should be used for plasticizing and
tackifying. Applications include adhesives, inks, carpet back-
ings, and rubber, paper and fabric coatings.

Doversperse A-1:
   % Solids: 65
   Poise/25C: 48
   % Cl: 45
   Appearance: Cream White
   SG/25C: 1.60

Doversperse 3:
   % Solids: 66.5
   Poise/25C: 150-300
   % Cl: 45
   Appearance: Cream White
   SG/25C: 1.54

DSM Melapur: MELAPUR Flame Retardants:

Melapur 200:
    CAS Nr.: 218768-84-4
    Formula: $(C_3H_8N_6)n$ $(HPO_3)m$
    Chemical Name: Melamine-polyphosphate
    Appearance: Fine, white, crystalline powder
Specifications:
    Nitrogen Content: 42-44 wt.%
    Phosphor Content: 12-14 wt.%
    Water Content: max 0.2 wt.%
    Particle Size: D50%: max. 10um
              D98%: max. 25um
Other Properties:
    Water Solubility 20C: <0.01 g/100 ml
    pH value (saturated sol., 20C): approx. 5
    Specific gravity: 1.85 g/cm3
    Bulk density: approx. 3.00 kg/m3
Applications: Flame retardant for polyamide 66

Application Data:
    Melapur 200 is a halogen-free flame retardant for glass filled
polyamide 66. As a white powder it opens up new colour schemes
for flame retardant GFR PA 66. The mechanism is based on intum-
escence. On exposure to fire a strong and stable char is formed
which protects the polymer and prevents further flame propagation. Formulations with 25-30 wt% of Melapur 200 in GFR PA66
achieve UL 94 V-O rating at a thickness of 1.6 mm. The amount of
flame retardant has to be established according to polymer,
glass fibre and application. The good balance between intumes-
cence and thermal stability of Melapur 200 provides a processing
window of up to 320C. Excellent health and safety data offer
additional handling and processing advantages.
Extrusion Data:
    Extruder: ZSK 30.33 W+P, twin screw, synchronous
    Screw speed: 100-150 rpm
    L/D: 33
    Feeding equipment: Gravimetric scale K'tron for feeding of
        Melapur 200 and polyamide
        Gravimetric scale Engelhardt KDE-FS 100 E, forced feeding
        of glass fibre.
    Residence time FR: 30-50 sec.
    Output: 10-12 kg/h (torque 85%)
    Vacuum: -0.8 bar
    PA 66: <0.10 wt.% H2O
Formulation:
    Polyamide 6.6: Bayer Durethane A 31/DSM Akulon S222: 45-50wt%
    Glass fibre: Vetrotex-P327/OCF 173X10C: 25wt%
    Flame Retardant: Melapur 200: 25-30wt%

DSM Melapur: MELAPUR Flame Retardants (Continued):

Melapur MC50: Melamine Cyanurate:
    EINECS Nr.: 2535757
    CAS Nr.: [37640-57-6]
    Formula: C6H9N9O3
    Appearance and Identity: fine, white, crystalline powder
                              IR spektrum
Specifications:
    Melamine Cyanurate: min. 99.5 wt%
    Water content: max. 0.2 wt%
    Excess Melamine: max. 0.3 wt%
    Excess Cyanuric acid: max. 0.2 wt%
    Particle size: D99%: max. 50um
Other Properties:
    Water solubility 20C: 0.001g/100ml
    pH Value (saturated sol., 20C): 5-6
    Mol Weight: 255.2
    Specific gravity: 1.7 g/cm3
    Bulk density: 300-400 kg/m3

Applications:
    Flame Retardant for polyamides, epoxy resins, polyurethanes,
polyolefines, polyester.
Other applications include: Lubricant, Adhesion agent

Melapur MP: Melamine Phosphate:
    EINECS Nr.: 243-601-5
    CAS Nr. [20208-95-1]
    Formula: C3H9N6O4P
    Appearance and Properties: Fine, white powder, not hygroscopic
and non-flammable
Specifications:
    Molar ratio Melamine/Phosphonic acid: 0.95-1.10
    Melamine Phosphate: min. 98 wt%
    Melamine content: min. 54 wt%
    Phosphoric acid content: min. 41 wt%
    Water content: max. 2 wt%
    Particle size: 99 wt% <50um
Other Properties:
    Mol. Weight: 224.12
    Water solubility (20C): <0.25g/100 ml
    Bulk density: 300-400 kg/m3
    Phosphorus content: min 12 wt%
    Nitrogen content: min 36 wt%
    pH value (saturated sol., 22C): 2.5-3.5
    Specific gravity: 1.74 g/cm3

Applications:
    Flame Retardant for thermoset and thermoplastic formulations,
intumescent systems

Eastern Color & Chemical Co.: ECCOGARD Flame Retardants:

Eccogard A-5:
A durable flame retardant for urethanes, polyesters, poly-
styrene, polypropylene and SBR.
Properties:
Appearance: A water white slightly viscous liquid
Active Content: 100%
Solubility: Soluble in mineral oil, organic solvents and
           esters. Not water soluble.
Density: 11.8 lbs/gallon
General Comments:
Eccogard A-5 is a durable flame retardant for many non-aqueous
polymer systems. When properly compounded Eccogard A-5 will
supply a durable flame retardancy. The applied finish is
durable to water, cleaners and solvents.
Eccogard A-5 is non-toxic and may be safely handled.
Recommended applications will provide flame retardancy
which will give satisfactory results to NFPA 701 and MVSS 302
flame testing conditions.
Application:
Polyurethanes: 0.5-5% on total weight
Polyesters: 2-5%
Polystyrene: 5-10%
Polypropylene: 2-10%
Exact amount of Eccogard A-5 will depend upon polymer used.

Eccogard A-7:
A durable flame retardant based on chlorinated phosphonates
for urethanes, polyesters, polystyrene, polypropylene and SBR.
Properties:
Appearance: A water white slightly viscous liquid
Active Content: 100%
Solubility: Soluble in mineral oil, organic solvents and
           esters. Not water soluble
Density: 11.8 lbs/gallon
General Comments:
Eccogard A-7 is a durable flame retardant for many non-aqueous
polymer systems. When properly compounded, Eccogard A-7 will
supply a durable flame retardancy. The applied finish is durable
to water, cleaners and solvents.
Eccogard A-7 is non-toxic and may be safely handled by workers
Recommended applications will provide flame retardancy which
will give satisfactory results to NFPA 701 and 305 Flame Testing
conditions.
Application:
Polyurethanes: 0.5-5% on total weight
Polyesters: 2-5%
Polystyrene: 5-10%
Polypropylene: 2-10%
Exact amount of Eccogard A-7 will depend upon the polymer used

**Eastern Color & Chemical Co.: ECCOGARD Flame Retardants
(Continued):**

Eccogard A-10 Base:
A durable flame retardant for urethanes, polyesters, poly-
styrene, polypropylene and SBR based on polycyclic organo-
phosphonate.
Properties:
Appearance: A water white very viscous liquid
Active Content: 100%
Solubility: Soluble in mineral oil, organic solvents and
esters, dispersible in water.
Density: 11.8 lbs/gallon
Viscosity cps @ 72F: 1,250,000 typical
               @ 140F: 6,500 typical
               @ 212F: 400 typical
Volatiles @ 400F: nil
          @ 500F: 1-2
          @ 575F: 6-7
          @ 650F: 10-11
General Comments:
Eccogard A-10 Base is a durable flame retardant for many non-
aqueous polymer systems. When properly compounded, Eccogard A-10
Base will supply a durable flame retardancy. The applied finish
is durable to water, cleaners and solvents.
Eccogard A-10 Base is non-toxic and may be safely handled.
Recommended applications will provide flame retardancy which
will give satisfactory results to NFPA 701 and MVSS 302 Flame
Testing conditions.
Application:
Polyurethanes: 0.5-5% on total weight
Polyesters: 2-5%
Polystyrene: 5-10%
Polypropylene: 2-10%
Exact amount of Eccogard A-10 Base will depend upon the polym-
er used.

Eccogard A-12:
A highly effective amino-phosphate based flame retardant for
nylon, polyesters, polystyrene, polypropylene, SBR and urethanes.
Properties:
Appearance: White powder
Active Content: 100%
Density: 10.1 lbs/gallon
Bulk Density: 5.25 lbs/gallon
General Comments:
Eccogard A-12 is a highly effective flame retardant for poly-
olefins and other polymer systems. When properly compounded
Eccogard A-12 will supply a durable flame retardancy without
affecting melt-flow properties.
Eccogard A-12 is non-toxic and may be safely handled.
Satisfactory results to NFPA 701 and MVSS 302 Flame Testing
Application:
Polypropylene: 15-40%   Low Density Polyethylene: 25-40%
Nylon 6,66: 15-35%   High Density Polyethylene: 30-45%

Ferro Corp.: FYARESTOR Flame Retardant Additives:

Fyarestor flame retardant additives are designed for highly effective cost/performance utilization in a wide variety of applications. These additives are bromochlorinated short chain hydrocarbons, that is, they contain both bromine and chlorine halogens as the flame retarding agents. These products allow the lower-cost introduction of halogens into the intended flame retarding applications and make use of the flame-retarding synergism between bromine and chlorine.

In addition, the additives act as secondary plasticizers in, for example, coating and film applications where flexibility of the finished product is required. Because of this, the amounts of plasticizers normally used can be reduced to reduce cost savings while achieving the desired flame retardancy.

In order to improve the flame retarding action of these Fyarestor additives, antimony trioxide can be employed as a flame retarding synergist in these applications which tolerate the use of a powdered additive. The use of this metal oxide will reduce the flame retarding cost. Also, the Fyarestor products can be emulsified for use in water-based systems.

Typical Properties:

Fyarestor 100:
  % Bromine: 20
  % Chlorine: 40
  Viscosity (SUS @ 210F): 160
  Color (Gardner): 2
  Appearance: Clear Liquid

Fyarestor 104:
  % Bromine: 30.5
  % Chlorine: 32
  Viscosity (SUS @ 210F): 2
  Color (Gardner): 2
  Appearance: Clear Liquid

Fyarestor 102:
  % Bromine: 22
  % Chlorine: 31
  Viscosity (SUS @ 210F): 65
  Color (Gardner): 1
  Appearance: Clear Liquid

Fyarestor 330B Flame Retardant Additive:

Fyarestor 330B flame retardant additive is a unique and proprietary liquid additive which produces excellent cost/ performance results in a variety of applications. This product is completely water soluble and therefore can be applied to a significant number of applications which require this character-istic. Fyarestor 330B contains and utilizes both halogen (bromine) and phosphorus flame retarding systems. Although these two systems produce flame retardancy by two different mechanisms, there is a provable synergistic action between them, that is, the combination of these two systems produces more efficient flame retardancy then does either one by itself. Also, it is not necessary to employ metal oxide synergists, such as antimony trioxide, with Fyarestor 330B.

Typical Properties:

  Solids, Weight %: 33%
  Water, Weight %: 67%
  Viscosity, Centipoise @ 25C: 3.5

  On dry basis:
  Bromine, Wt. %: 55%
  Phosphorus, Wt. %: 8%

Ferro Corp.: PYRO-CHEK Flame Retardant Additives:

Pyro-Chek 68PB:

Pyro-Chek 68PB flame retardant additive is a brominated polystyrene polymer, and is therefore both thermoplastic and polymeric in nature. It is unique in that this polymeric additive
will fuse at higher compounding temperatures and melt-blend
into the host polymer system. This property results in both
excellent dispersibility and subsequent molding performance.
The good thermal stability of Pyro-Chek 68PB allows it to be
melt-blended into most engineering resin systems. Since this
additive itself is a polymer, migration, or blooming, is eliminated in the final compounded plastic. This unusual polymeric
additive allows a good balance of properties to be achieved
in the compounded plastic, as well as providing unusually
good flow and excellent moldability.
   Bromine Content (%): 66.0 min.

Pyro-Chek 68PBC:

Pyro-Chek 68PBC flame retardant additive is a brominated
polystyrene polymer, and is therefore both thermoplastic and
polymeric in nature. It is unique in that this polymeric additive will fuse at higher compounding temperatures and melt-blend
into the host polymer system. This compacted grade of additive
consists of irregular granules about 1/8 inch in size. The
granular form results in a radical reduction in the amount of
environmental dusting when the additive is used. Since this
additive itself is a polymer, migration, or blooming, is eliminated in the final compounded plastic. This unusual polymeric
additive allows a good balance of properties in the compounded
plastic, as well as providing unusually good flow and excellent
moldability.
   Bromine Content (%): 66.0 min.

Pyro-Chek 68PBG:

Pyro-Chek 68PBG flame retardant additive is a brominated
polystyrene polymer, and is therefore both thermoplastic and
polymeric in nature. This finely-ground form of the additive
is intended for use in applications where the compounding and
molding temperatures are not high enough to fuse, or melt,
the polymeric additive. The intended temperature use for this
ground, or small particle size, additive is below 230C (450F).
In this range, the additive will remain a filler-type additive
and the small particle size will help maintain the physical
properties of the fully compounded plastic. The unusual polymeric additive allows a good balance of properties to be
achieved in the compounded plastic, as well as providing
unusually good flow and excellent moldability.
   Bromine Content (%): 66.0 min.

Ferro Corp.: PYRO-CHEK & BROMOKLOR Flame Retardant Additives

Pyro-Chek 60PBC:
Pyro-Chek 60PBC flame retardant additive is a brominated
polystyrene polymer, and is therefore both thermoplastic and
polymeric in nature. It is unique in that this polymeric
additive will fuse at higher compounding temperatures and melt-
blend into the host polymer system. This compacted grade of
additive consists of irregular granules about 1/8 inch of size.
The granular form results in a radical reduction in the environmental dusting when the additive is used. Since this additive
itself is a polymer, migration, or blooming, is eliminated
in the final compounded plastic. This dibrominated polymer
exhibits a lower glass transition temperature than the other
Pyro-Chek additives, which are tribrominated. At a given temperature, it therefore provides better dispersibility and flow
characteristics, but less bromine content.
Bromine Content (%): 58.0 min.

Bromoklor Flame Retardant Additives:
Bromoklor flame retardant additives are designed for highly
effective cost/performance utilization in a wide variety of
applications. These additives are bromo-chlorinated short
chain hydrocarbons, that is, they contain both bromine and
chlorine halogens as the flame retarding agents. This allows
the lower-cost introduction of halogens into the intended flame
retarding applications and makes use of the flame retarding
synergism between bromine and chlorine.
In addition, the additives act as secondary plasticizers in,
for example, coating and film applications where flexibility
of the finished product is required. Because of this, the
amounts of plasticizers normally used can be reduced to produce
cost savings while achieving the desired flame retardancy.
In order to improve the flame retarding action of these
Bromoklor additives, antimony trioxide can be employed as a
flame retarding synergist in these applications which tolerate
the use of a powdered additive. Also, the Bromoklor additives
can be emulsified easily for use in water-based systems.

Typical Properties:

Bromoklor 50
- % Bromine: 30
- % Chlorine: 21
- Free HCl, ppm: 4
- Color (Gardner): 1
- Appearance: Clear Liquid

Bromoklor 70:
- % Bromine: 30
- % Chlorine: 37
- Free HCl, ppm: 6
- Color (Gardner): 2
- Appearance: Clear Liquid

Focus Chemical Corp.: Akzo Nobel Chemicals: FYROL Flame Retardants:

Fyrol FR-2:
   Chemical Name: tri-(1,3-dichloroisopropyl) phosphate
   Physical Appearance: Colorless liquid
   Phosphorus Content Wt%: 7.1
   Chlorine Content Wt%: 49.0
Applications:
   * Flexible polyurethane foam
     - excellent processing, thermal and hydrolytic stability
     - particularly suitable for achieving Ca. Bulletin 117
       and automotive FMVSS-302
     - low fogging in the DIN 75201 test for automotive foams
     - adjunct FR for HR-melamine foams.
   * Epoxy resins and phenolics
   * Unsaturated polyester resin

Fyrol FR-2LV:
   Chemical Name: modified tri(1,3-dichloroisopropyl) phosphate
   Physical Appearance: Colorless liquid
   Phosphorus Content Wt%: 7.3
   Chlorine Content Wt%: 47.0
Applications:
   * Low viscosity version of Fyrol FR-2 with same FR performance
     but less tendency to crystallize.
     - not suitable for foams where fogging is important

Fyrol 38:
   Chemical Name: modified tri(1,3-dichloroisopropyl) phosphate
   Physical Appearance: Pale amber liquid
   Phosphorus Content Wt%: 7.0
   Chlorine Content Wt%: 49.0
Applications:
   * "Scorch" stabilized version of Fyrol FR-2 designed to reduce
     discoloration caused by high exotherms of high water content
     formulations.
   * Flexible polyurethane foam
     - same application as Fyrol FR-2

Fyrol CEF:
   Chemical Name: tri(2-chloroethyl) phosphate
   Physical Appearance: Colorless liquid
   Phosphorus Content Wt%: 10.8
   Chlorine Content Wt%: 36.7
Applications:
   * Rigid and flexible polyurethane foams and coatings.
   * Bonded polyurethane foams.
   * Epoxy resins and phenolics.
   * Adjunct FR for HR-melamine foams.
   * Polymethyl methacrylates.
   * Unsaturated polyester resins.

Focus Chemical Corp.: Akzo Nobel Chemicals: FYROL Flame Retardants (Continued):

Fyrol PCF:
    Chemical Name: tri(2-chloroisopropyl) phosphate
    Physical Appearance: Colorless liquid
    Phosphorus Content Wt%: 9.5
    Chlorine Content Wt%: 32.5
Applications:
    * Rigid polyurethane foam systems
      - excellent thermal and hydrolytic stability (A or B
        components).
      - particularly suitable for ASTM E-84 (Class II)
    * Bonded polyurethane foams.
    * Unsaturated polyester resins and phenolics
      - low viscosity (low temperature use).

Fyrol 25:
    Chemical Name: chloroalkyl phosphate/oligomeric phosphonate
    Physical Appearance: Dark gray liquid
    Phosphorus Content Wt%: 10.2
    Chlorine Content Wt%: 37.0
Applications:
    * Flexible polyether urethane foams.
      - low scorch in high water content formulations.
      - excellent efficiency.
      - low fogging in the DIN 75201 test for automotive foams.

Fyrol 99:
    Chemical Name: oligomeric chloroalkyl phosphate
    Physical Appearance: Colorless liquid
    Phosphorus Content Wt%: 14.0
    Chlorine Content Wt%: 26.0
Applications:
    * Rigid polyurethane foam systems.
      - efficient FR to achieve UL 94 HF-1.
    * Unsaturated polyester resins.
    * Thermosets (epoxy systems and phenolics).
    * Acrylic latices and molding compounds

Fyrol PBR:
    Chemical Name: brominated flame retardant based on pentabromo
                    diphenyl ether
    Physical Appearance: Dark amber liquid
    Phosphorus Content Wt%: 2.2
    Chlorine Content Wt%: 51.5% Br
Applications:
    * Flexible polyether polyurethane foam
      - excellent where very low scorch/discoloration is a prime
        consideration in foam slab production.

Focus Chemical Corp.: Akzo Nobel Chemicals: FYROL Flame Retardants (Continued):

Fyrol DMMP:
  Chemical Name: Dimethyl methyl phosphonate
  Physical Appearance: Colorless liquid
  Phosphorus Content Wt%: 25.0
Applications:
  * Non-halogen flame retardant for use in rigid polyurethane
    foam systems
    - very effective and efficient.
    - effective in combination with brominated flame retardants.
    - suitable for pentane blown films.
  * Unsaturated polyester resins.
    - suitable in combination with alumina trihydrate for molded
      parts (reduces viscosity of resin).
Fyrol 6:
  Chemical Name: Diethyl N,N bis(2-hydroxyethyl) amino methyl
                 phosphonate
  Physical Name: Amber liquid
  Phosphorus Content Wt%: 12.0
Applications:
  * Reactive non-halogen flame retardant for rigid polyurethane
    foams.
    - incorporated into the foam structure.
    - excellent for spray, pour-in-place systems and slab foam.
  * Adjunct flame retardant with other FR'S e.g. Fyrol DMMP.
Fyrol 51:
  Chemical Name: oligomeric phosphonate
  Physical Appearance: Pale yellow liquid
  Phosphorus Content Wt%: 20.5
Applications:
  * Cellulosic products
    - aqueous latex FR backcoating systems for cotton fabrics.
    - paper air filters using phenolic resin coatings.
  * Phenolic resins
  * Paints (latex emulsions)
  * Polyesters (polyurethane terephthalate)
Fyrol 58:
  Chemical Name: Fyrol 51 and an organic solvent
  Physical Appearance: Pale yellow liquid
  Phosphorus Content Wt%: 17.1
Applications:
  * Improved processing over Fyrol 51 where films are applied
    followed by drying
    - useful for cellulosic automotive and industrial air filters
Fyrol 42:
  Chemical Name: proprietary customized additive for flame
                 lamination
  Physical Appearance: Colorless liquid
  Phosphorus Content Wt%: 5.3
Applications:
  *Enhances the flame bonding efficiency of flexible polyether
   urethane foam.

Focus Chemical Corp.: Akzo Nobel Chemicals: FYROL Melamine Derivatives:

Fyrol MP:
   Chemical Name: Melamine phosphate 1,3,5-Triazine-2,4,6 tria-
              mine, phosphate
   Molecular Weight: 224
   Appearance: White powder
   % Phosphorus: 13.8
   % Nitrogen: 38
Applications:
   * Flame retardant for:
     * Polyolefins   * Elastomers and   *Engineering resins
   * Intumescent coatings for various substrates
   * Rigid polyurethane foams
   * Coating wood particle boards and paper

Fyrol MC:
   Chemical Name: Melamine cyanurate 1,3,5-Triazine-2,4,6
              (1H, 3H, 5H) trione, compound with 1,3,5-
              triazine-2,4,6 triamine
   Molecular Weight: 255
   Appearance: White crystalline powder
   % Nitrogen: 49
Applications:
   * Flame retarding polyamide molding resins
   * Polyolefins   * Polyesters   *Epoxy resins and   *Rubber
   * Adjuvant flame retardant for intumescent coatings and paints
   * Higher temperature processing due to high thermal stability
   * Additive in lubricants
   * Adhesive agent for metal coatings and lacquers on plastics

Focus Chemical Corp.: Bayer AG: Triethyl Phosphate (Phosphoric Acid Triethylester):

Description:
   Chemical formula: C6H15O4P
   Molar mass: 182 g/mole
   Clear, colourless, low-viscosity liquid with a slight, ester-like odour

Specification:
   TEP content: % by wt.: min. 99.5
   Acid value: mg KOH/g: max. 0.05
   Water content: % by wt.: max. 0.2
   Hazen colour value: max. 20
   Refractive index n20D: 1.405-1.407

Typical Analysis:
   TEP content: % by wt.: 99.9
   Acid value: mg KOH/g: 0.005
   Water content: % by wt.: 0.03
   Hazen colour value: 5
   Density at 20C: g/cm3: 1.069
   Viscosity at 20C: mPa-s: 1.7
   Refractive index n20D: 1.406
   Setting point: C: -56
   Boiling point at 5 mbar: C: 80
   Flash point (open cup): C: 130

Applications:
   TEP is an extremely strong and versatile aprotic solvent. Since TEP is highly polar, it dissolves a large number of polar materials which are often not readily soluble in other solvents. TEP also has flame-retardant properties.
   Thus, TEP can be used as a solvent as well as a flame-retardant in materials such as plastics based on cellulose, polyester resins and polyurethanes.
   This product also serves as a versatile solvent in organic syntheses.
   TEP is a solvating and desensitizing agent for organic peroxides.
   It is used as a catalyst for synthesizing ketene in the production of acetic anhydride by the Wacker process.

GE Silicones: SFR100 Silicone Fluid:

SFR100 silicone fluid is a high viscosity silicone containing a combination of a linear silicone fluid, and a silicone resin which is soluble in the fluid. The resulting mixture is clear and colorless. SFR100 has been tailored for polyolefins and thermoplastics use to impart flame retardancy. When compounded properly with a Group II A metal organic salt and other ingredients, varying degrees of flame retardance can be obtained.

Key Performance Properties:
* Non-halogen flame retardant
* Replaces antimony/halogen flame retardant system
* Improves impact strength even at low temperatures
* Excellent electricals (non-conductive or excellent electrical properties)
* Improved processability
* Improves melt and mold flow

Typical Product Data:
Silicone Content: 100%
Viscosity, at 25C: 200,000-900,000 cps
Specific Gravity: 1.00
Flash Point* C: 202
Appearance: Clear, Colorless Liquid
Shelf Life, Unopened Drums: 6 months
Density: 8.7 lbs/gal avg.
Density: 1.0 g/cc avg.
*Pensky-Martens

Instructions for Use:
Best results are obtained by compounding in a twin screw extruder using a specific screw element sequence, barrel temperature and screw speed. Due to high viscosity and tackiness of SFR fluid, it is difficult to prepare a permanently free flowing pre-mix of the SFR100 fluid, polypropylene and other dry ingredients. Low shear mixing (Hobart mixer) will produce a powder which is free flowing for several days at room temperature. High shear (Henschel or dough mixer) produces a tacky granular material, suitable for a Banbury type mixer, but not a twin screw extruder. It is most convenient to add SFR100 fluid separately into the extruder - allowing the dry ingredients to be mixed upstream, with SFR100 fluid being added downstream, after the polypropylene has melted when viscosities are similar making a better mix.

Banbury Compounding:
Due to high viscosity of SFR100 fluid, better results are obtained if all dry ingredients are fluxed and melted prior to addition of SFR100 fluid. After the addition of SFR100 fluid, melt blend should be completed.

Extrusion Compounding:
The twin screw extrusion process was found to be very effective for compounding flame retardant addiitves into polypropylene.

Great Lakes Chemical Corp.: Flame Retardants:

Dibromostyrene and Derivatives:
Great Lakes DBS:
  Dibromostyrene   CAS No. 125904-11-2
  Formula Weight: 261.9   Viscosity: 4 cps @ 25C
  Bromine Content: 59.0%   Specific Gravity: 1.8

Great Lakes PDBS-80:
  Poly (dibromostyrene)   CAS No. 148993-99-1
  Formula Weight: 80,000   Melting Range C: 220-240
  Bromine Content: 59.0%   Specific Gravity: 1.9

Great Lakes GPP-36:
  Polypropylene-dibromostyrene graft copolymer
  Bromine Content: 36.0%   Melting Range C: 160-175
  Graft copolymer of polypropylene/DBS Specific Gravity: 1.3
  CAS No. 171091-06-8

Great Lakes GPP-39:
  Proprietary derivative of GPP-36   Melting Range C: 150-170
  Bromine Content: 39.0%   Specific Gravity: 1.3
  Graft copolymer of polypropylene/DBS

Tetrabromophthalic Anhydride and Derivatives:
Great Lakes PHT4:
  Tetrabromophthalic anhydride CAS No. 632-79-1
  Formula Weight: 463.7   Melting Range C: 274-277
  Bromine Content: 68.2%   Specific Gravity: 2.9

Great Lakes PHT4-DIOL:
  Tetrabromophthalate diol   CAS No. 77098-07-8
  Formula Weight: 627.9   Viscosity: 100,000 cps @ 25C
  Bromine Content: 46.0%   Specific Gravity: 1.9

Great Lakes DP-45:
  Tetrabromophthalate ester   CAS No. 26040-51-7
  Formula Weight: 706.1   Viscosity: 1800 cps @ 25C
  Bromine Content: 45.1%   Specific Gravity: 1.6

Firemaster BZ-54:
  Tetrabromobenzoate ester   Viscosity: 500 cps @ 25C
  Bromate Content: 54.1%   Specific Gravity: 1.7
  Proprietary

Great Lakes Chemical Corp.: Flame Retardants (Continued):

Hexabromocyclododecane and Derivatives:
Great Lakes CD-75P:
   Hexabromocyclododecane   CAS No. 3194-55-6
   Formula Weight: 641.7   Melting Range C: 180-192
   Bromine Content: 74.7%   Specific Gravity: 2.1

Great Lakes SP-75:
   Stabilized hexabromocyclododecane
   Bromine Content: 72.0%   Melting Range C: 187-192
   CAS No. 3194-55-6   Specific Gravity: 2.1

Tetrabromobisphenol A and Derivatives:
Great Lakes BA-59P:
   Tetrabromobisphenol A   CAS No. 79-94-7
   Formula Weight: 543.7   Melting Range C: 179-182
   Bromine Content: 58.8%   Specific Gravity: 2.2

Great Lakes PE-68:
   Tetrabromobisphenol A bis (2,3-dibromopropyl ether)
   Formula Weight: 943.6   Melting Range C: 106-120
   Bromine Content: 67.7%   Specific Gravity: 2.2
   CAS No. 21850-44-2

Great Lakes BE-51:
   Tetrabromobisphenol A bis (allyl ether)
   Formula Weight: 624.0   Melting Range C: 115-120
   Bromine Content: 51.2%   Specific Gravity: 1.8
   CAS No. 25327-89-3

Great Lakes BC-52:
   Phenoxy-terminated carbonate oligomer of Tetrabromobisphenol A
   Formula Weight: 2,500   Proprietary
   Bromine Content: 51.3%   Melting Range C: 180-210
   CAS No. 94334-64-2   Specific Gravity: 2.2

Great Lakes BC-52HP:
   Phenoxy-terminated carbonate oligomer of Tetrabromobisphenol A
   Formula Weight: 5,300   Proprietary
   Bromine Content: 53.9%   Melting Range C: 210-240
   CAS No. 94334-64-2   Specific Gravity: 2.2

Great Lakes BC-58:
   Phenoxy-terminated carbonate oligomer of Tetrabromobisphenol A
   Formula Weight: 3,500   Proprietary
   Bromine Content: 58.7%   Melting Range C: 200-230
   CAS No. 71342-77-3   Specific Gravity: 2.2

Great Lakes Chemical Corp.: Flame Retardants (Continued):

Halogenated Phosphate Esters:
Firemaster HP-36:

Halogenated phosphate ester   Proprietary
Formula Weight: 416.5   CAS No. 125997-20-8
Halogen Content: 44.5%   Viscosity: 240 cps @ 25C
Phosphorus Content: 7.5%   Specific Gravity: 1.6

Firemaster 836:

Halogenated phosphate ester   Proprietary
Formula Weight: 416.5   Viscosity: 240 cps @ 25C
Halogen Content: 44.5%   Specific Gravity: 1.6
Phosphorus Content: 7.5%

Firemaster 642:

Halogenated phosphate ester   Proprietary
Halogen Content: 49.2%   Viscosity: 400 cps @ 25C
Phosphorus Content: 6.5%   Specific Gravity: 1.7

Intumescent Flame Retardants:
Great Lakes NH-1197:

Intumescent flame retardant   CAS No. 5301-78-0
Halogen Content: 0%   Melting Range C: 195-202
Phosphorus Content: 17.0%   Specific Gravity: 1.7

Great Lakes NH-1511:

Intumescent flame retardant   Proprietary
Halogen Content: 0%   Melting Range C: 211-215
Phosphorus Content: 15.0%   Specific Gravity: 2.6

Brominated Diphenyl Oxides:
Great Lakes DE-83R:

Decabromodiphenyl oxide   CAS No. 1163-19-5
Formula Weight: 959.2   Melting Range C: 300-310
Bromine Content: 83.3%   Specific Gravity: 3.3

Great Lakes DE-79:

Octabromodiphenyl oxide   CAS No. 32536-52-0
Formula Weight: 801.4   Melting Range C: 85-89
Bromine Content: 79.8%   Specific Gravity: 2.8

Great Lakes DE-71:

Pentabromodiphenyl oxide   CAS No. 32534-81-9
Formula Weight: 564.7   Viscosity: >200,000 cps @ 25C
Bromine Content: 70.8%   Specific Gravity: 2.3

Great Lakes Chemical Corp.: Flame Retardants (Continued):

Brominated Diphenyl Oxides (Continued):
Great Lakes DE-60F Special:
    Pentabromodiphenyl oxide blend
    Bromine Content: 52.0%   Viscosity: 2000 cps @ 25C
    Proprietary DE-71 Blend   Specific Gravity: 1.9

Great Lakes DE-61:
    Pentabromodiphenyl oxide blend
    Bromine Content: 51.0%   Viscosity: 2000 cps @ 25C
    Proprietary DE-71 Blend   Specific Gravity: 1.9

Great Lakes DE-62:
    Pentabromodiphenyl oxide blend
    Bromine Content: 51.0%   Viscosity: 2000 cps @ 25C
    Proprietary DE-71 Blend   Specific Gravity: 1.9

Tribromophenol and Derivatives:
Great Lakes PH-73:
    2,4,6-Tribromophenol   CAS No. 118-79-6
    Formula Weight: 330.8   Melting Range C: 91-95
    Bromine Content: 72.5%   Specific Gravity: 2.2

Great Lakes PHE-65:
    Tribromophenyl allyl ether CAS No. 3278-89-5
    Formula Weight: 371.0   Melting Range C: 74-76
    Bromine Content: 64.2%   Specific Gravity: 2.1

Great Lakes PO-64P:
    Poly (dibromophenylene oxide) CAS No. 69882-11-7
    Formula Weight: 6,000   Softening Range C: 210-240
    Bromine Content: 62.0%   Specific Gravity: 2.3

Great Lakes FF-680:
    bis (Tribromophenoxy) ethane CAS No. 37853-59-1
    Formula Weight: 687.6   Melting Range C: 223-228
    Bromine Content: 70.0%   Specific Gravity: 2.6

H & C Industries, Inc.: FIREX 300C Flame Retardant:

Firex 300C is a flame retardant concentrate specially designed to impart good flame retardancy to both Homopolymer and Copolymer Polypropylene. It can be used in Blow Molding, Extrusion, Injection Molding, Structural Foam and Film Manufacturing. At the loading levels usually required, Firex 300C does not significantly affect the mechanical and electrical properties of the resin.

Firex 300C has excellent thermal stability and is able to withstand prolonged high temperature processing without any degradation or causing significant discoloration to the resin. Firex 300C is non-volatile, non-blooming, chemically inert, and has a very high degree of UV light stability.

UL Component Program Recognition:
Firex 300C is recognized under the Component Program of Underwriters Laboratories Inc. for UL 94V-2 Flammability standard covering all colors and all melt flows of Homopolymer Polypropylene in a minimum 1:10 let-down ratio.

Typical Properties:
    Appearance: 1/8 inch off-white pellets
    Composition: Carrier Resin, 40%
                 Flame Retardant Additives, 60%
    Melting Point (carrier resin): 176C
    Specific Gravity: 2.18
    Decomposition Temperature: Above 300C

Recommended Loading Levels:
Due to varieties of Polypropylene and different applications, it is difficult to predict the exact loading levels required to meet specific flammability standards. H&C usually recommends loading level at 10% (by weight of total compound) as the starting formulation point and then, based upon results of flammability tests, gradually move upward or downward to achieve maximum cost savings and meet the required flammability standards.

Toxicology:
Firex 300C has been found non-toxic to rats on single-dose ingestion, and is non-irritating to skin and eyes.

H & C Industries, Inc.: FIREX 400C Flame Retardant Masterbatch:

Firex 400C is a newly developed highly effective & easy-dispersing flame retardant masterbatch for use in Polyolefin & EVA based systems such as films, foams, sheets, wire & cable insulation, elastomers, etc. It is suggested for use in Lamination, Extrusion, Injection Molding, and Film/Foam Manufacturing.

Firex 400C has excellent thermal stability and can withstand processing temperature up to 400C. Also, at the loading levels usually required, Firex 400C will not significantly affect the physical & electrical properties of the finished products.

Typical Properties:
  Appearance: White pellets
  Composition: Flame retardants-60%
                 Polyethylene resin-40%
  Melting Point (resin): 120C
  Specific Gravity: 2.60

Packaging:
  Net 100 kgs in fiber drum with inner PE liner

Recommended Loading Levels:
  Coating for Canvas products*   12-15%* (by weight)
  Low Density Polyethylene system   9-15%
  High Density Polyethylene system   12-18%
  Thin Film & Foam   7-10%
(*Apply on LDPE coating for passing CPAI-84 Vertical flame test)

J.M. Huber Corp.: SOLEM Alumina Trihydrate: Thermosets Product Selecton Guide:

Alumina Trihydrate is the largest volume flame retardant used in the world. Also known as ATH and hydrated alumina, it is technically aluminum hydroxide, with the chemical formula $Al(OH)3$.

ONYX Elite 100:
  Cast Polymer: Gel Coated

Onyx Elite 200:
  Cast Polymer: Gel Coated

Onyx Elite 255:
  Cast Polymer: Gel Coated/Solid Surface

Onyx Elite 332:
  Composites: Continuous Panel/RTM (Resin Transfer Molding)

Onyx Elite 335:
  Cast Polymer: Gel Coated/Solid Surface
  Composites: Continuous Panel/Electrical Laminates/General
    Molded Products/SMC/BMC

Onyx Elite 336:
  Cast Polymer: Encapsulating/Potting
  Composites: Electrical Laminates/General Molded Products/
    SMC/BMC

Onyx Elite 431:
  Cast Polymer: Gel Coated/Encapsulating/Potting
  Composites: RTM (Resin Transfer Molding)

Onyx Elite 432:
  Cast Polymer: Solid Surface/Encapsulating/Potting
  Composites: Continuous Panel/Electrical Laminates/General
    Molded Products/RTM (Resin Transfer Molding)/SMC/BMC

Onyx 632:
  Cast Polymer: Solid Surface

SB-36:
  Cast Polymer: Encapsulating/Potting
  Composites: General Molded Products/SMC/BMC

SB-122:
  Cast Polymer: Encapsulating/Potting
  Composites: Continuous Panel/Electrical Laminates/General
    Molded Products/Pultrusion/RTM (Resin Transfer Molding)/
    SMC/BMC/Spray-Up/Hand Lay-Up

J.M. Huber Corp.: SOLEM Alumina Trihydrate: Thermosets Product Selection Guide (Continued):

SB-136:
   Cast Polymer: Encapsulating/Potting
   Composites: General Molded Products/Spray-Up/Hand Lay-Up

SB-332:
   Composites: Continuous Panel/General Molded Products/Pultru-
      sion/RTM (Resin Transfer Molding)/Spray-Up/Hand Lay-Up

SB-336:
   Cast Polymer: Encapsulating/Potting
   Composites: Continuous Panel/Electrical Laminates/Filament
      Winding/General Molded Products/Phenolic Molding/RTM
      (Resin Transfer Molding)/SMC/BMC/Spray-Up/Hand Lay-Up

SB-432:
   Cast Polymer: Encapsulating/Potting
   Composites: Continuous Panel/Electrical Laminates/Filament
      Winding/General Molded Products/Phenolic Molding/Pultru-
      sion/RTM (Resin Transfer Molding)/SMC/BMC/Tooling Applica-
      tions

SB-632:
   Composites: Electrical Laminates/Filament Winding/Polyurethane
      Elastomer/RTM (Resin Transfer Molding)/SMC/BMC/Spray-Up/
      Hand Lay-Up

FRE/HYFIL:
   Composites: Spray-Up/Hand Lay-Up

MICRAL 885:
   Composites: Polyurethane Elastomer

Micral 932:
   Composites: Electrical Laminates/General Molded Products/
      Polyurethane Elastomer/RTM (Resin Transfer Molding)/
      SMC/BMC

Micral 1440:
   Cast Polymer: Encapsulating/Potting
   Composites: Electrical Laminates/General Molded Products/
      RTM (Resin Transfer Molding)/SMC/BMC/Spray-Up/Hand Lay-Up

Micral 9400:
   Composites: Electrical Laminates/General Molded Products/
      Pultrusion/SMC/BMC

Laurel Industries: DECHLORANE PLUS Flame Retardant:

More of the benefits you are looking for:
A free-flowing white powder, Dechlorane Plus contains 65
percent chlorine in a cycloaliphatic compound, ideal for impart-
ing flame retardant properties to thermoplastics, thermosets
and elastomers. As the only stable chlorinated flame retardant
on the market, Dechlorane Plus is usually combined with anti-
mony oxide (Sb2O3), but in some resins can be used with other
synergists.
CAS registry number 13560-88-9
Dechlorane Plus is available in three particle sizes:
* Dechlorane Plus 515: 1-25 microns (avg. 9)
* Dechlorane Plus 25: 1-10 microns (avg. 5)
* Dechlorane Plus 35: 0.5-5 microns (avg. 2)
A variety of synergist options:
* Does not require the use of antimony oxide in nylons.
* Mixed synergists can be used with nylon 6, nylon 66
  and epoxies.
* Combinations of various zinc synergists, alone or with
  iron oxides, can be used.
* Results in lower cost formulation and improved properties.
UV stability reduces discoloration:
* Aliphatic structure does not absorb UV rays.
* Minimal discoloration after prolonged aging.
* Ideal for applications where appearance is important.
Thermal stability means easier processing:
* Operating temperatures up to 320C allow greater ease of
  processing in a wide variety of polymers.
Inert filler will not disrupt operations:
*Non-plasticizing, non-reactive, thermally stable.
*Imparts hydrolytic stability and excellent electrical prop-
  erties, even with exposure to moisture.
Easy color coding and matching:
* Fine white powder allows easy color coding and pastel
  matching.
* Helps maintain the original color of computer housings and
  other machines.
Non-blooming, non-bleeding
* Insoluble filler offers good resistance to blooming (bleed-
  ing) in finished products.
High CTI performance:
* Comparative Tracking Index (CTI) values of greater than 400
  volts for a number of polymer systems, including flame
  retarded nylons can be achieved.
Low smoke generation:
* Generates 95 percent less smoke than commercial bromine-
  based additives.
* Important for plastics compounds and others containing
  polyolefin polymer systems.
Extremely cost effective:
* Extremely low density (1.8) vs. most brominated flame
  retardants (2.2-3.5).
* Exceptionally cost effective on a volume/cost basis.

Laurel Industries: FIRESHIELD Flame Retardants:

FireShield Antimony Oxides:
The particle size of any non-melting plastic additive can affect a polymer's physical properties, such as tensile and elongation properties, impact strength and opacity. Laurel's highly controlled manufacturing processes allow the selection of specific particle size distributions tailored to your specific requirements.

General Purpose Grades:

FireShield H:
The average particle size of FireShield H is 1.0 to 1.8 microns. Because of this smaller particle size, it has a less degrading effect on the physical properties of plastics. It also has a higher tint strength than FireShield L.

FireShield L:
With an average particle size of 2.5 to 3.5 microns, Fire-Shield L has a lower tint strength and minimizing effect on pigmentation. This reduces the amount of colored pigments required, especially when making brightly colored PVC products.

Specialty Grades:

ULTRAFINE II:
UltraFine II has an average particle size of 0.2 to 0.4 microns, the smallest available on the market. It is recommended for applications where a minimum loss of physical properties is required. Flame-retarded ABS, for example, can be prepared with only a minimal loss of original impact strength. UltraFine II is highly effective in liquid systems such as unsaturated polyesters, epoxies, polyurethanes, phenolics and flame-retardant textile treatments. Its tinting strength is slightly greater than FireShield H.

FireShield HPM Series:
FireShield HPM Series High Purity antimony oxide is used as a catalyst, chemical intermediate and flame retardant in special-ty applications requiring low levels of lead, arsenic and other trace metals. FireShield HPM is available in two grades. The 1.0 to 1.8 micron average particle size is suitable for many sensi-tive electronic applications, while the 0.2 to 0.4 submicron grade is used in liquid systems such as epoxies, urethanes, adhesives, paints, solvents and colloids where dispersibility or surface area is important.

PETCAT Series:
The Petcat Series are high-purity products with superior ethylene glycol and HCl solubility designed specifically for use as a catalyst in polyester (PET) production. Petcat is available in both a standard grade (1.0 to 1.8 microns) and a submicron grade (0.2 to 0.4 microns).

Laurel Industries: LSFR Flame Retardant Additive:

LSFR:
Physical Properties:
   Color: White
   Appearance: Free-flowing powder
   Bulk Density: 1.0g/cc or 63 lbs./ft3 approx.
   Specific Gravity: 4.20
   Average Particle Size: 1.1u

   Reduces smoke generation up to 35 percent

   High flame retardant efficiency
   As a flame retardant, laboratory tests show that LSFR additive performs exceptionally well to inhibit flammability.

   Increased thermal stability
   In tests measuring thermal stability, the compound with LSFR showed a dramatic reduction in color shift compared to the other compounds.

   Improved ultraviolet stability
   LSFR also improves the ultraviolet stability of polymers

   Superior long-term durability
   In addition, LSFR retained its tensile strength and elongation properties on a comparable basis with the other samples in tests measuring accelerated aging.

   Easy processing...readily dispersible
   In laboratory tests on a two-roll mill, compounding of LSFR additive was much easier than compounding of the other synergists because LSFR acted as a processing aid, providing external lubrication.
   Furthermore, the relatively small particle size of LSFR makes it easily dispersible, improving its effective synergism.

   LSFR additive, a proprietary flame retardant synergist, meets the increasing demand for PVC and other polymer products that are not only fire resistant, but smoke resistant as well. Further, LSFR enhances the physical properties of polymers in six additional ways, making it one of the most cost-effective flame retardant additives available. In addition, it offers low specific gravity and bulk density, and no compromise to your compound's physical integrity.

Laurel Industries: THERMOGUARD and PYRONIL 45 Flame Retardants:

Laurel Industries' flame retardants are based on Thermoguard antimony-halogen and Pyronil brominated plasticizer technology and used to flame retard a variety of plastics and other materials. Synergistic action occurs when Thermoguard antimony is combined with a halogen which continues to be a very cost effective flame retardant technology for most polymers.

Thermoguard antimony oxide is available at a flame retardant purity grade typically 99.5% antimony oxide. The standard particle size grade is Thermoguard S at 1.0-1.8 microns. Thermoguard L has a larger particle size of 2.0-3.2 microns, which is used to minimize the pigmentation effect.

Thermoguard FR sodium antimonate is used instead of antimony oxide to prevent premature degradation in certain polymers such as PET when used with a halogenated flame retardant to achieve flame retardant performance.

Low dusting grades of Thermoguard antimony oxide and CPA are achieved by wetting with a variety of liquids such as mineral oil, water, DOP, DIDP and liquid chlorinated paraffins.

Thermoguard 505, decabromodiphenyl oxide (DBBO) is a high bromine halogen source used in combination with Thermoguard antimony oxide for a low cost effective synergist package.

Thermoguard 8218 is in pelletized concentrate form and contains 82% decabromodiphenyl oxide (DBBO). The halogen source, used in combination with Thermoguard antimony oxide, is used to complete the antimony-halogen synergist package.

Pyronil 45 is a flame retardant liquid which contains 45% bromine. An added benefit is the restoration of physical properties normally lost when formulating with mineral filled or glass fiber compositions.

Pyronil 63, a melt processable solid flame retardant containing 63% bromine, continues under development.

Thermoguard flame retardant pelletized concentrates/masterbatches which elmininate dusting, improve processing and provide easier handling are available.

Thermoguard antimony oxide at 80% or 90% with polyethylene as a carrier.

Thermoguard FR sodium antimonate up to 80% with either EEA, polyethylene and other polymers as a carrier.

Thermoguard 8218 contains 82% decabromodiphenyl oxide (Thermoguard 505-DBBO) with polyethylene as a carrier.

Thermoguard 243-S contains a brominated epoxy resin and Thermoguard S antimony oxide for flame retarding thermosets and thermoplastics.

Martin Marietta Magnesia Specialties: MAGCHEM Magnesium Oxides:

MagChem 40 Light Burned Technical Grade Magnesium Oxide:
  MagChem 40 is a high purity technical grade of magnesium oxide
processed from magnesium-rich brine. This fine white powder has
a very high reactivity and a low bulk density.
  MagChem 40 is well suited for many rubber formulations, part-
icularly neoprene. It finds wide use as a filler, anti-caking
agent, and pigment extender. MagChem 40 is extremely efficient
in chemical processes where ease of conversion is a factor. It
is also used in the production of oil additives and the desilic-
ation of water.
Composition:
  Magnesium oxide (MgO), % ignited basis: 98.2
Physical Properties: Typical:
  Bulk density, loose, lb/ft3 (g/cm3): 22 (0.35)
  Mean particle size, microns: 5
  Surface area, m2/g: 45
  Activity index, seconds: 9
  Screen size, % passing 325 mesh, wet: 99.5

MagChem 50 Light Burned Technical Grade Magnesium Oxide:
  MagChem 50 is a high purity technical grade of magnesium
oxide processed from magnesium-rich brine. This fine white
powder has a very high reactivity index, low bulk density,
and excellent flow properties.
  MagChem 50 is well suited for chemical reactions where high
reactivity and rapid conversion to magnesium hydroxide are
required because of its very high surface area. Its high purity
and controlled activity make it an excellent magnesium source
for the manufacture of many magnesium compounds. It is used in
rubber compounding, plastic thickening, filtrate clarification,
odor removal, and a variety of selective adsorptions.
Composition:
  Magnesium oxide (MgO), % ignited basis: 98.2
Physical Properties: Typical:
  Bulk density, loose, lb/ft3 (g/cm3): 18 (0.29)
  Mean particle size, microns: 5
  Surface area, m2/g: 65
  Activity index, seconds: 8
  Screen size, % passing 100 mesh: 100.0
  Screen size, % passing 325 mesh, wet: 99.5

Martin Marietta Magnesia Specialties: MAGCHEM Magnesium Oxides (Continued):

MagChem 50M Light Burned Technical Grade Magnesium Oxide:

MagChem 50M is a high purity technical grade of magnesium oxide processed from a magnesium-rich brine. This fine white powder has a very high reactivity index, low bulk density, and excellent flow properties.

MagChem 50M is well suited for chemical reactions where high reactivity is required because of its high surface area and extra fine particle size. It is recommended for use in rubber and plastic compounding.

Composition:
Magnesium oxide (Mg), % ignited basis: 98.0

Physical Properties: Typical:
Bulk density, loose, lb/ft3 (g/cm3): 18 (0.29)
Mean particle size, microns: 0.8
Surface area, m2/g: 65
Activity index, seconds: 8
Screen size, % passing 100 mesh: 100.0
Screen size, % passing 325 mesh, wet: 99.5

MagChem 60 Light Burned Technical Grade Magnesium Oxide:

MagChem 60 is a high purity technical grade of magnesium oxide processed from magnesium-rich brine. This fine white powder has a very high reactivity index, low bulk density, and excellent flow properties.

MagChem 60 is well suited for chemical reactions where high reactivity and rapid conversion to magnesium hydroxide are required because of its very high surface area. Its high purity and controlled activity make it an excellent magnesium source for the manufacture of many magnesium compounds. It is used in rubber compounding, plastic thickening, filtrate clarification, odor removal, and a variety of selective adsorptions.

Composition:
Magnesium oxide (MgO), % ignited basis: 98.0

Physical Properties: Typical:
Bulk density, loose, lb/ft3 (g/cm3): 18 (0.29)
Mean particle size, microns: 5
Surface area, m2/g: 80
Activity index, seconds: 6.5
Screen size, % passing 100 mesh: 100.0
Screen size, % passing 325 mesh, wet: 99.5

Martin Marietta Magnesia Specialties: MAGSHIELD Magnesium Hydroxide for Flame-retardancy and Smoke Suppression

MagShield provides flame retardance through five mechanisms:

1   MagShield produces more char than ATH resulting in increased effectiveness and less smoke.
2   MagShield dilutes the amount of fuel available to sustain combustion.
3   MagShield, during combustion, generates highly reflective magnesium oxide coating which deflects the flame's heat away from the polymer.
4   MagShield contains 31% bound water which is released beginning at 330C and blankets the flame to limit oxygen available for combustion.
5   MagShield absorbs 17% more heat from the combustion zone than ATH to reduce the prospect of continued burning.

The MagShield Advantage

MagShield allows processing at a temperature up to 100C higher than ATH. An increase of as little as 10 to 30C can significantly improve processing efficiency by increasing extrusion speeds and decreasing mold filling time.

MagShield is cost competitive with ATH for many applications making MagShield the performance and economical choice in flame retardants.

MagShield is less abrasive than ATH, resulting in longer mixing and compounding equipment life. This is demonstrated by MagShield's Mohs' Hardness of 2.5 vs. 3.0 for ATH.

MagShield exhibits low toxicity, [LD50 (oral/rat) >5,000 mg/kg] and is non-corrosive compared to halogen or phosphorus containing compounds.

MagShield S:

Standard Grade Magnesium Hydroxide for Flame Retardant Applications.

MagShield S magnesium hydroxide has a typical median particle size of 4.9 microns and is designed for flame retardant applications where the properties of this metal hydrate provide cost benefit advantages in the flame retardant compound.

Physical Properties (Typical):

Magnesium Hydroxide (Mg(OH)2), %: 98.7
Loss on Drying, %: 0.4
Loss on Ignition, %: 30.9
Median Particle Size, microns: 4.9
Surface Area, m2/g: 11
Specific Gravity: 2.36
Loose Bulk Density, g/ml: 0.40
Mohs Hardness: 2.5
TAPPI Brightness, %: 95

**Martin Marietta Magnesia Specialties: MAGSHIELD Magnesium
Hydroxide for Flame-retardancy and Smoke Suppression
(Continued):**

Magshield M:
    Medium Fine Grade Magnesium Hydroxide for Flame-retardant
Applications
    MagShield M magnesium hydroxide has a typical median particle
size range of 1 to 2.2 microns and is designed for flame-retardant applications where the properties of this metal hydrate
provide cost benefit advantages in the flame-retardant compound.
Physical Properties (Typical):
    Magnesium Hydroxide, ($Mg(OH)_2$), %: 98.7
    Loss on Drying, %: 0.4
    Loss on Ignition, %: 30.9
    Median Particle Size, microns: 1-2.2
    Surface Area, m2/g: 11
    Specific Gravity: 2.36
    Loose Bulk Density, g/ml: 0.40
    Mohs Hardness: 2.50
    TAPPI Brightness, %: 95

MagShield UF:
    Ultra-fine Grade Magnesium Hydroxide for Flame Retardant
Applications
    MagShield UF magnesium hydroxide has a typical median part-
icle size under 1 micron and is designed for critical applications such as flame retarded wire and cable insulating compounds
Physical Properties (Typical):
    Magnesium Hydroxide ($Mg(OH)_2$), %: 98.7
    Loss on Drying, %: 0.4
    Loss on Ignition, %: 30.9
    Median Particle Size, microns: 0.9
    Surface Area, m2/g: 12
    Specific Gravity: 2.36
    Loose Bulk Density, g/ml: 0.40
    Mohs Hardness: 2.5
    Hunter Color: 96

MagShield:
    Stearate Treated Magnesium Hydroxide for Flame Retardant
Polyolefin and PVC Applications
    MagShield S, and MagShield UF magnesium hydroxides are avail-
able with specific stearate coatings to enhance their performance in thermoplastic resins. The treatments can improve dispersion, reduce mixing requirements, improve physical properties
and allow increased loading.
Currently available grades:
    MagShield SNB5: 0.5% stearate
    MagShield UFNB10: 1.0% stearate

Nyacol Nano Technologies, Inc.: Colloidal Antimony Pentoxide Flame Retardant Additives:

Nyacol Nano Technologies, Inc. offers colloidal antimony pentoxide in various forms for use as synergists with halogenated flame retardants.

Advantages:
* Better penetration of the substrate.
* Less pigmenting or whitening effect for deep mass true colors.
* Easier handling and processing; liquid dispersions will not clog spray guns.
* High flame retardant efficiency for minimal added weight or change in hand.
* Greater cost effectiveness and improved color matching in mass tone colors.
* Translucency for applications such as coatings, films, and laminates.
* Suspension stability in low viscosity systems.
* Maintenance of physical properties in thin films and fine denier fibers.

Nyacol Organic Dispersions:
Organic colloidal dispersions find use in the plastics, coatings, and adhesives industries where water cannot be added.

AB40:
% Oxide: 40
Organic Base: Unsaturated Polyester
End Use: Polyester laminates, high temperature cure

AP50:
% Oxide: 50
Organic Base: Non-reactive high M.W. tertiary amine
End Use: Epoxy resins, Ketone solutions

APE1540:
% Oxide: 40
Organic Base: Unsaturated polyester
End Use: Polyester laminates, room temperature cure

Nyacol Nano Technologies, Inc.: Colloidal Antimony Pentoxide Flame Retardant Additives (Continued):

BurnEx Colloidal Antimony Pentoxide Powder:
BurnEx Powders:
Aqueous dispersions of colloidal antimony pentoxide are spray-dried to make antimony pentoxide powders. These powders will not tint the product because they are 10 to 40 micron agglomerates of nano-sized powders. In situations where Nyacol available resin carriers are not compatible with your system, Nyacol regular powders are preferred.

BurnEx Plus A1588LP:
End Use: Epoxies

BurnEx Plus A1590:
End Use: Epoxies

BurnEx ZTA:
End Use: Vinyl, ABS, HIPS, Polypropylene

BurnEx Nano-Dispersible Powders:
These powders of colloidal antimony pentoxide readily disperse to nanometer-sized particles in various solvents and polymers. Nyacol Nano Technologies, Inc. currently offers three nano-dispersible powders.

BurnEx A1582 disperses in water and has found use in vinyl and some thermoplastic polyesters.
BurnEx ADP480 is compatible with and disperses in non-polar hydrocarbons such as hexane and toluene, and in thermoplastics such as polypropylene.
BurnEx ADP494 disperses in polar compounds such as acetone, acetonitrile, and MEK. It is used for solvent extended epoxy, solution spun fibers and thermoplastics such as ABS.

BurnEx A1582:
End Use: Vinyl, Polyester
Compatibility: Water
BurnEx ADP480:
End Use: Polyolefins, HIPS
Compatibility: Non-Polar Solvents
BurnEx ADP494:
End Use: Solvent Epoxy, Fibers, ABS
Compatibility: Polar Solvents

Nyacol Nano Technologies, Inc.: Colloidal Antimony Pentoxide Flame Retardant Additives (Continued):

Polymer Concentrate:
BurnEx 2000:
BurnEx 2000 polypropylene flame retardant concentrates are made from nano-dispersible antimony pentoxide and an organic bromine compound. Polypropylene resins of different Melt Flow Index (MFI) are used for different applications. Nyacol Nano Technologies, Inc. can design other concentrates or work with your compounder to formulate a custom product for you. They have a fully equipped compounding laboratory at their Ashland, MA facility.

BurnEx 2000-10:
 % Active FR: 25
 Base: PP MFI 4
 End Uses: Blow Molding, Film

BurnEx 2000-20:
 % Active FR: 25
 Base: PP MFI 20
 End Uses: Fiber

Application:
Usage Levels:
The usage level of the various grades depends on the material being treated, the test procedure, the ratio of halogen to antimony oxide, and other ingredients being used. It is not possible to predict flame retardance performance; therefore, experimental evaluation is required. In textile, paper, and plastics applications, typical use levels of antimony pentoxide are 1.5 to 8%, with a halogen use level range of 5 to 20%. Typical ratios of halogen to antimony are 5:1 to 2:1.

Polymer Additives Group: CHARMAX AOM & MO Fire Retardant &
Smoke Suppressant Additives:

Charmax Ammonium Octamolybdate C-AOM:
    Charmax Ammonium Octamolybdate is an ultrafine white to off-
white powder of exceptionally high purity used as a flame retard-
ant synergist and smoke suppressant in polymers.
Typical Properties: Physical:
    Specific Gravity: 3.18
    Bulk Density: 35 lb/ft3
    Mean Particle Size: 1.5-2.5 microns
    99.9% less than: 15 microns
    Solubility in water: 5g/100 ml
    Decomposition Temp.: >480F
    Loss on Ignition: 8.29%
Chemical:
    Formula: $(NH_4)4Mo8O26$
    Analysis (theoretical): Mo 61.10%
                            NH3: 5.43%
Applications:
    AOM is designed to help formulators achieve reduced levels of
smoke with high levels of flame retardancy in rigid and flexible
PVC and PVC alloys and adhesives. Specific applications have also
been developed in urethanes, elastomers and other polymers.
Products made with C-AOM are used in transportation, construction
and wire & cable markets, where stringent smoke and flammability
standards, such as UL 910 and ASTM E84, must be met. Typical
uses include jacketing and insulation for plenum and riser;
profiles, wall coverings and upholstery for high risk buildings;
and extrusions for subways and aircraft interiors.
Charmax Molybdic Oxide C-MO:
    Charmax MO Molybdic Oxide is a high purity, free-flowing,
pale blue-gray powder utilized in polymers primarily as a
smoke suppressant.
Typical Properties: Physical:
    Average Particle Size (microns): 2.5
    99.9% Less Than (microns): 32
    Bulk Density (lb/ft3): 15
    Specific Gravity: 4.7
    Loss on Ignition at 600C (%): <3.0
Chemical:
    Analysis (theory): Mo 66.65%
Applications:
    C-MO is designed to help formulators achieve reduced levels
of smoke with high levels of flame retardancy in rigid and
flexible PVC and PVC alloys and adhesives. Specific applications
have also been developed in urethanes, elastomers and other
polymers. Products made with C-MO are used in transportation,
construction and wire & cable markets, where stringent smoke
and flammability standards, such as UL 910 and ASTM E84, must be
met. Typical uses include jacketing and insulation for plenum
and riser; profiles, wall coverings and upholstery for high
risk buildings; and extrusions for subways and aircraft interiors. For color sensitive applications the customer should use
Charmax Ammonium Octamolybdate (C-AOM) which is a white material.

Polymer Additives Group: CHARMAX Low Smoke:

Charmax Low Smoke C-LS-Family:
Charmax LS family of flame retardants and smoke suppressants
for polymers are ultrafine white powders of proprietary
inorganic complexes.

Typical Properties: Physical:
    Color: White to off white
    Specific Gravity: 3.20
    Mean Diameter: 2-3 microns
    99.9% less than: 20 microns
    Decomposition Temp.: >550F

Applications:
Charmax LS additives are designed to help the formulator
achieve reduced levels of smoke with high levels of flame
retardancy in halogenated and other polymers. They are particularly effective in PVC based wire and cable, sheet and film,
and coatings.

Charmax LS ZHS Zinc Hydroxystannate:
The Charmax LS ZHS is a zinc hydroxystannate of high purity
and fine particle size and is used as a flame retardant and smoke
suppressant in polymers.

Typical Properties: Physical:
    Color: white to off-white
    Specific Gravity: 3.30
    Mean Particle Size: 1.5-2.0 microns
    99% less than: 25 microns
    Decomposition Temp.: >375F
    Free Moisture: <0.9%

Applications:
Charmax LS ZHS is designed to help the formulator achieve high
levels of flame retardancy with reduced levels of smoke and
carbon monoxide evolution in halogenated and other polymers.
They are particularly effective in PVC based wire & cable, rigid
film and coatings, and halogenated polyesters. They can also be
used as charformers in non-halogenated PE/EVA and epoxy systems.

Charmax LS ZST Zinc Stannate:
The Charmax LS ZST is a zinc stannate of high purity and fine
particle size and is used as a flame retardant and smoke suppressant in polymers.

Typical Properties: Physical:
    Color: white to off-white
    Specific Gravity: 3.90
    Mean Particle Size: 1.5-2.5 microns
    99% less than: 25 microns
    Decomposition Temp.: >570F
    Free Moisture: <0.9%

Applications:
Charmax LS ZST is designed to help the formulator achieve
high levels of flame retardancy with reduced levels of smoke
and carbon monoxide evolution in halogenated and other polymers.

Polymer Additives Group: CHARMAX FR Flame Retardants

Charmax FR Z4S, Z8S, Z20S: C-FR Family:
The Charmax FR family of flame retardants and smoke suppressants for polymers are ultrafine white powders of proprietary inorganic complexes.
Typical Properties: Physical:
  Color: white to off-white
  Specific Gravity: 3.60
  Mean Particle Size: 2-3 microns
  99% Less than: 25 microns
  Decomposition Temp.: >550F
Applications:
Charmax FR additives are designed to help the formulator achieve high levels of flame retardancy with reduced levels of smoke in halogenated and other polymers. They are particularly effective in PVC based wire & cable, sheet and film, and coatings.

Charmax FR INT-3: C-FR Family:
Charmax FR flame retardants and smoke suppressants for polymers are ultrafine white crystalline powders of proprietary inorganic complexes.
Typical Properties: Physical:
  Color: white
  Specific Gravity: 4.50
  Mean Particle Size (microns): 2-3
  99% Less Than (microns): 25
  Decomposition Temp. F: >550
Applications:
Charmax FR INT-3 is designed to help the formulator achieve high levels of flame retardancy and antimony trioxide replacement in halogenated polymers including PVC. C-FR INT-3 also reduces smoke emissions compared with antimony trioxide and may provide lower and more consistent tinting strength than some commercial grades of antimony trioxide.

Charmax FR Antimony Oxide (low tint):
Charmax AO antimony oxide (trioxide) is a white, odorless, crystalline powder. Insoluble in water, it is soluble in concentrated hydrochloric and sulfuric acids, and strong alkalies.
Typical Properties: Physical:
  Average Particle Size (microns): 2.3
  % Through 325 Mesh: 99.9
  Specific Gravity: 5.5
Chemical:
  Formula: Sb_2O_3
  Antimony Trioxide: 99.5%
Applications:
Flame Retardant Synergist-Antimony oxide in combination with a halogen source constitutes an extremely effective flame retardant for use in broad range of polymers.
Chlorine or bromine are the halogens most often employed, and may be present in the polymer chain.

Polymer Additives Group: CHARMAX Zinc Borates:

Charmax Zinc Borates: ZB 200 Series:
The Charmax ZB Series of Zinc Borates are white, ultrafine, free-flowing powders.
Typical Properties: Physical:
    Average particle size: 2-3 microns
    99.9% less than: 15 microns
    Solubility in water: 0.1g/100ml
    Specific Gravity: 2.70
    Bulk Density: 20 lb/ft3
    Oil Absorption: 31g/100g
    Refractive Index: 1.57
    TGA: 1% wt loss: 390F
Chemical:
    Formula: $2ZnO-2B_2O_3-3H_2O$
    ZnO (theor.): 45.7%
    B2O3 (theor.): 39.1%
    H2O (theor.): 15.2%
Applications:
The Charmax ZB 200 series zinc borates are effective flame retardant synergists and smoke suppressants in halogenated polymers, particularly PVC, where they allow the partial or complete replacement of antimony oxide. They act as strong charformers in the condensed phase, inhibiting combustion and smoke formation by isolating the substrate from atmospheric oxygen. Applications include calendared sheet, compounded thermoplastics, plastisols, elestomers and thermosets. Typical uses for Charmax ZB 200 include jacketing and insulation for plenum and riser cables, carpet tiles, wall coverings and belting.

Charmax Zinc Borates: ZB 400 Series:
The Charmax ZB Series of Zinc Borates are ultrafine, white flame retardant/smoke suppressant powders that can also be used for reduced afterglow and arc/track.
Typical Properties-Physical:
    Average Particle Size: 2-3 microns
    99.9% less than: 15 microns
    Solubility in water: 0.2g/100ml
    Specific Gravity: 2.74
    Bulk Density: 30 lb/ft3
    Refractive Index: 1.59
    TGA: 1% wt loss: 500F
Chemical:
    Formula: $4ZnO-6B_2O_3-7H_2O$
    ZnO (theor.): 37.4%
    B2O3 (theor.): 48.1%
    H2O (theor.): 14.5%
Applications:
The ZB 400 Series of Charmax Zinc Borates retain their water of hydration at high enough temperatures to process polyamides, polypropylene, and thermoplastic elastomers where they are also used as afterglow suppressants and anti-arcing agents.

Polymer Additives Group: HYDRAMAX Magnesium Hydroxides:

Hydramax fire retardant and smoke suppressant magnesium hydroxides include fine and intermediate particle size FR additive grades. These free flowing white powders provide a cost effective way to flame retard and smoke suppress a variety of polymers.

HM-93C:
Physical:
    Brightness: 85+
    Specific Gravity: 2.4
    Bulk Density (lbs/ft3): 53
    Mean Particle Size (microns): 25-30
    Retained on 325 mesh (%): n/a
    Processing Temperature (C): >320
Chemical:
    Magnesium as Mg(OH)3: 93.6
    Calcium as CaCO3: 3.10

HM-93:
Physical:
    Brightness: 88+
    Bulk Density (lbs/ft3): 43
    Mean Particle Size (microns): 10-15

HM-955:
Physical:
    Brightness: 95+
    Bulk Density (lbs/ft3): n/a
    Mean Particle Size (microns): 4-9
Chemical:
    Magnesium as Mg(OH)3: 98.5
    Calcium as CaCO3: 0.5

HM-933:
Physical:
    Brightness: 95+
    Bulk Density (lbs/ft3): n/a
    Mean Particle Size (microns): 3-4
Chemical:
    Magnesium as Mg(OH)3: 98.5
    Calcium as CaCO3: 0.5

HM-922:
Physical:
    Brightness: 95+
    Bulk Density (lbs/ft3): n/a
    Mean Particle Size (microns): 2-3
Chemical:
    Magnesium as Mg(OH)3: 98.5
    Calcium as CaCO3: 0.5

These FR products are used in various thermoplastics including PVC and polyolefins, rubber, high temperature adhesives/coatings.

Polymer Additives Group: HYDRAMAX Magnesium Hydroxides
(Continued):

Hydramax HM-B8 & HM-B8-S:
  Hydramax Fire Retardant and Smoke Suppressant Magnesium based
materials include some of the finest particle size, highest
quality grades of FR additives available. These free flowing
white powders provide a cost effective way to flame retard and
smoke suppress engineering plastics, rubber, high temperature
adhesives, coatings and other high temperature polymer systems.
Typical Properties:
Physical:
    Median Particle Size: 1.1
    Top Size-99% Less than (microns): 8
    Free moisture (%): 0.5
    Loss on Ignition (%): 31.
    Surface Area (m2/g): 8
    Specific Gravity: 2.38
    Oil Absorption: 30
    Refractive Index: 1.58
    Hardness (Mohs): 2.5
    TAPPI Brightness: 94+
    Processing Temperature (C): 320
Chemical:
    MgO%: 65.0
    CaO%: 2.4
    Fe2O3%: 0.35
    Total Na2O%: 0.20
Applications:
  These FR products are used in flexible and rigid PVC, nitrile
rubbers, neoprene, polyolefins, EPDM, SBR, EPR, urethanes, EVA
copolymers, nylons and other engineering polymer systems where
magnesium hydroxides are used. For high loadings in engineering
resins HM-B8-S is preferred.

Hydramax HM-C9 & HM-C9-S:
Typical Properties:
Physical:
    Median Particle Size (microns): 1.1
    Top Size-99% less than (microns): 9.0
    Free moisture (%): 0.5
    Loss on Ignition (%): 54.0
    Surface Area (m2/g): 19
    Specific Gravity: 2.60
    Oil Absorption: 40
    Refractive Index: 1.6
    TAPPI Brightness: 95+
    Processing Temperature: 260
Chemical:
    MgO%: 39.7   Fe2O3%: 0.01
    CaO%: 6.9   Total Na2O%: 0.20
Applications:
  These FR products are used in cross-linked polyethylene,
polypropylene, PBT, and other engineering polymer systems where
magnesium calcium carbonate hydroxides are used.

Polymer Additives Group: HYDRAX Alumina Trihydrate ATH:

Hydrax alumina trihydrates include some of the finest particle size, highest quality grades of ATH available. These free flowing white powders provide a cost effective way to flame retard and smoke suppress plastics, rubber, paper, adhesives, coatings and other polymer systems.

Typical Properties: Coarse off-white ground products

H-312:
    Median Particle Size (microns): 12
    Top Size-99% Less than (microns): 55
    Free moisture (%): 0.3
    Loss on Ignition (%): 34.6
    Bulk Density/loose (lbs/ft): 45
    Specific Gravity: 2.42
    Refractive Index: 1.57
    Hardness (Mohs): 3
    TAPPI Brightness: 88+

H-314:
    Median Particle Size (microns): 14
    Top Size-99% Less than (microns): 55
    Bulk Density/loose (lbs/ft): 50

H-216:
    Median Particle Size (microns): 16
    Top Size-99% Less than (microns): 59
    Bulk Density/loose (lbs/ft): 56

H-218:
    Median Particle Size (microns): 18
    Top size-99% Less than (microns): 61
    Bulk Density/loose (lbs/ft): 58

H-120:
    Median Particle Size (microns): 20
    Top size-99% Less than (microns): 62
    Bulk Density/loose (lbs/ft): 62

H-136:
    Median Particle Size (microns): 36
    Top size-99% Less than (microns): 69
    Bulk Density/loose (lbs/ft): 68

Applications:
    These ATH products are used in flexible and rigid PVC, nitrile rubbers, neoprene, polyolefins, EPDM, SBR, EPR, latexes urethanes, EVA copolymers, unsaturated polyesters and other systems.

Polymer Additives Group: HYDRAX Alumina Trihydrate ATH (Continued):

H-470:
Median Particle Size (microns): 7.0
Top Size-99% Less than (microns): 44
Free moisture (%): 0.4
Loss on Ignition (%): 34.6
Bulk Density/loose (g/cm3): 0.64
Bulk Density/packed (g/cm3): 1.34
Surface Area (m2/g): 2
Specific Gravity: 2.42
Oil Absorption: 16
Refractive Index: 1.57
Hardness (Mohs): 3
TAPPI Brightness: 88+

H-490:
Median Particle Size (microns): 9.0
Top Size-99% Less than (microns): 44
Bulk Density/loose (g/cm3): 0.76
Bulk Density/packed (g/cm3): 1.34

H-550:
Median Particle Size (microns): 5.0
Top Size-99% Less than (microns): 16
Bulk Density/loose (g/cm3): 0.6
Bulk Density/packed (g/cm3): 1.1

H-636:
Median Particle Size (microns): 3.6
Top Size-99% Less than (microns): 11
Bulk Density/loose (g/cm3): 0.56
Bulk Density/packed (g/cm3): 0.95

H-826:
Median Particle Size (microns): 2.6
Top Size-99% Less than (microns): 9.0
Bulk Density/loose (g/cm3): 0.42
Bulk Density/packed (g/cm3): 0.82

H-910:
Median Particle Size (microns): 1.2
Top Size-99% Less than (microns): 6.0
Bulk Density/loose (g/cm3): 0.20
Bulk Density/packed (g/cm3): 0.40

Applications:
These ATH products are used in flexible and rigid PVC, nitrile rubbers, neoprene, polyolefins, EPDM, SBR, EPR, latexes, urethanes, EVA copolymers, unsaturated polyesters and other systems.

Sherwin-Williams Co.: KEMGARD Flame Retardant/Smoke Suppressant:

Kemgard products are cost effective flame retardants/smoke suppressant additives for polymeric systems. They are based on patented surface treatment technology.

Kemgard product chemistries include zinc molybdate, calcium zinc molybdate and zinc oxide/phosphate complexes.

Kemgard Features & Benefits:
* Low Tint Strength
* Low Specific Gravity
* Promotes Char Formation
* Non-Hazardous White Powder
* Low cost/Highly efficient smoke suppressant
* Compatible with other halogenated and non-halogenated flame retardants.

Kemgard Helps Meet the Following Performance Test:

ASTM E 84	ASTM D 2843	JIS 1321
ASTM E 662	ASTM D 2859	NFPA 263
ASTM E 906	ASTM D 2863	UL 94
ASTM E 1354	ASTM D 4100	UL 910

Kemgard 911A:
A high efficiency calcium zinc molybdate compound
Kemgard 911B:
A high efficiency basic zinc molybdate compound
Kemgard 911C:
A high efficiency zinc molybdate-magnesium silicate compound
Kemgard 981:
A zinc phosphate/zinc oxide compound
Kemgard 425:
A low cost calcium zinc molybdate compound
Kemgard 501:
A calcium molybdate compound for polymers destabilized by zinc at elevated processing temperatures.

Sybron Chemicals Inc.: SYBRON/TANATEX Flame Retardants:

Sybron/Tanatex offers a complete line of durable and non-durable flame retardant chemicals. Each has been specially formulated to produce superior flame retardant properties on the specified fiber.

Non-Durable Flame Retardants:
    FLAMEGARD AFP
    FlameGard DSH*
    FlameGard SR*
    FlameGard STS*
    PYROLUX AS-5M*
    Pyrolux PCS*
    Pyrolux WRB*

Durable Flame Retardants:
    FlameGard 908*
    FlameGard PE Conc.*
    Resin CNX Special

    *Registered in the state of California

Durable Flame Retardants:
    Polyester: FlameGard PE Conc.
    Nylon: FlameGard 908
          Resin CNX Special

FlameGard PE Conc.: Is recommended for use on 100% polyester
FlameGard 908: Is a thermosetting urea/thiourea formaldehyde resin type flame retardant producing a soft to moderate firm hand.
Resin CNX Special: A thermosetting urea/thiourea formaldehyde resin type flame retardant, which produces an extremely stiff hand.

Non-Durable Flame Retardants:
FlameGard DSH: Is as close to being a universal flame-retardant finish.
FlameGard STS: An effective low cost flame retardant effective on cotton and polyester/cotton fabrics.
FlameGard AFP: Has the least detrimental effect on water-repellents, while still being an extremely effective flame retardant
Pyrolux AS-5M: An effective flame retardant which produces a soft drapable hand
Pyrolux WRB: Recommended for use with specific water repellent finishes. Recommended mainly for cellulosics.
Pyrolux PCS: A semi-durable flame retardant designed for polyester/cotton fabrics.
FlameGard SR: Very soft hand, non-stiffening at high add-ons. Especialy designed for acetate, polyester and their blends.

3V Inc.: PLASTISAN B Flame Retardant:

C3H6N6-C3H3N3O3
1,3,5-Triazine-2,4,6 (1H, 3H, 5H)-trione, compd. with
1,3,5-Triazine-2,4,6-triamine (1:1)
or
Melamine cyanurate

Molecular weight: 255.2
CAS number: 37640-57-6
EINECS number: 2535757

Chemical and Physical Characteristics:
    Appearance: white powder
    Bulk density: approx. 0.25 g/cm3
    Solubility: practically insoluble in water and in most
               common solvents
    Particle size: min. 95% <10 microns
    Plastisan B is a thermally very stable compound. It possesses
neither a melt point, nor a softening range. Its initial decomposition and volatilization temperature is over 350C. Plastisan
B is a non-hygroscopic product.

Properties:
    Plastisan B is a compound belonging to the class of melamine
salts. Owing to its characteristics, it offers a wide range of
applications into plastic materials, as flame retardant. Plastisan B is compatible with various polymers.
Plastisan B offers the folllwing advantages:
    High efficiency: even at low addition percentages
    Thermal stability and low volatility: Owing to these char-
acteristics it is possible to formulate the product during
processing without loss of the additive through decomposition
or volatilization.
    Color: The addition of Plastisan B does not adversely affect
pigmentation of finished articles. It is recommended that tests
are carried out.
    Halogen free: The combustion gases are free from halogen or
hydrogen halide.

Applications:
    Plastisan B offers its best performances as flame retardant
    in:
    * Polyamides
    It is also effective in the following polymers:
    * Polyurethanes
    * Polyolefins
    * Polyesters
    Polyamides: in accordance with the method UL-94, it is poss-
ible to obtain a fire resistance of class V-2/V-1 or V-O, by
adding to the polymer (PA-6 or PA-6,6) 3-30% by wt. of Plastisan
B.

U.S. Borax Inc.: FIREBRAKE Fire Retardants:

Firebrake ZB:
An unique zinc borate combining the optimum effects of zinc
and boron oxides and water release for developing fire retardant
formulations processable up to 290C.

Firebrake ZB-XF:
A zinc borate offering the same properties as Firebrake ZB.
It can be used in applications where a smaller particle size is
of utmost importance.

Firebrake 415:
Having a higher zinc to boron ratio and much higher dehydra-
tion temperature than Firebrake ZB, this zinc borate is preferred
in applications where the processing temperatures can be up to
415C. Improved smoke reduction has also been noted with this zinc
borate in PVC.

Firebrake 500:
An anhydrous form of Firebrake ZB, especially designed for
applications where no water evolution is permitted while still
offering the performances associated with the optimum combination
of zinc and boron in Firebrake ZB.

Firebrake ZB:
Zinc Borate
Firebrake ZB
Firebrake ZB-Fine
CAS Number 138265-88-0

Applications:
Firebrake ZB is used as a flame retardant, smoke and afterglow
suppressant, and anti-arcing agent in polymer systems such as
polyvinyl chloride, nylon, epoxy, polyethylene, polypropylene,
polyesters, thermoplastic elastomers and rubbers.
Firebrake ZB-Fine is recommended for applications where maximum fire test performance is needed, and physical properties
such as film forming and adhesion are critical.

Chemical and Physical Properties:
Boric Oxide, B_2O_3: 48.05%
Zinc Oxide, ZnO: 37.45%
Water of Crystallization, H_2O: 14.50%
Anhydrous Equivalent: 85.50%

Characteristics (typical values):
Refractive Index: 1.58
Median Particle Size:
ZB: 7 microns (Sedigraph)
ZB-Fine: 3 microns (Sedigraph)

U.S. Borax, Inc.: FIREBRAKE Fire Retardants (Continued):

Firebrake ZB-XF:
Zinc Borate

Firebrake ZB-XF fire retardant is intended to replace Firebrake ZB-Fine for use in applications that require maximum fire test performance results and critical physical properties such as in thin film forming. In contrast to Firebrake ZB-Fine, the XF grade has no particles greater than 12 microns, as determined by Laser Diffraction technique.
Typical Chemical and Physical Properties:
Weight Loss at Elevated Temperatures: 0.8% at 290C (typical)
Particle Size: Typical median size 2 microns (micrometers)
Specific Gravity: 2.77

Firebrake 415:
Zinc Borate
CAS Number 1332-07-6
Applications:
Firebrake 415 can be used as a flame retardant and smoke suppressant in a variety of polymers. It is particularly useful in systems such as nylon that require high processing temperatures, where enhanced thermal stability has been noted in some cases. Firebrake 415 is also an excellent smoke suppressant for systems such as flexible PVC.
Chemical and Physical:
Theoretical Composition:
Boric Oxide, B_2O_3: 16.85%
Zinc Oxide, ZnO: 78.79%
Characteristics (typical values):
Refractive Index: 1.65
Median Particle Size: 5 microns (Laser Diffraction)

Firebrake 500:
Zinc Borate
CAS Number 1332-07-6
Applications:
Firebrake 500 can be used as a flame retardant and smoke suppressant in a variety of polymers including polyetherketone, polysulfone, fluoropolymer, polyester and nylon. The very significant beneficial effect of Firebrake 500 on rate of heat release is of special interest where this factor is important, as in aircraft applications.
Theoretical Composition:
Boric Oxide, B_2O_3: 56.20%
Zinc Oxide, ZnO: 43.80%
Characteristics (typical values):
Refractive Index: 1.58
Median Particle Size: 10 microns (Laser Diffraction)

Trade Name Index

Trade Name	Supplier
AEROSIL	Degussa
ALCAN	Alcan Chemicals
ALKANOX	Great Lakes Chemical
ALVINOX	3V Inc.
ALVIPACK	3V Inc.
AMSPEC	Amspec Chemical
AMSTAR	Amspec Chemical
ANOX	Great Lakes Chemical
ANTIBLAZE	Albright & Wilson Americas
ANTISTATICO	3V Inc.
APACIDER	Sangi America
AQUAFORTE	ADM Tronics Unlimited
ARCTIC MIST	Luzenac America
ARMOSTAT	Akzo Nobel
ASP	Engelhard
ATMOS	Witco
ATMUL	Witco
ATOMITE	ECC International
AZOFOAM	Biddle Sawyer
BENEFOS	Mayzo
BENNOX	Mayzo
BLO-FOAM	Rit-Chem
BROMOKLOR	Ferro
BUCA	Engelhard
BURN EX	Nyacol
CAMEL-CAL	ECC International
CAMEL-CARB	ECC International
CAMEL-FIL	ECC International
CAMEL-FINE	ECC International
CAMEL-TEX	ECC International
CAMEL-WITE	ECC International
CARB-O-FIL	Shamokin Filler
CATALPO	Engelhard
CHARMAX	Polymer Additives Group
CHARTWELL	Chartwell
CHEMSTAT	Chemax
CIMPACT	Luzenac America
CP FILLER	ECC Internatonal
CYANOX	Cytec Industries
CYASTAT	Cytec Industries
DECHLORANE PLUS	Laurel Industries
DICAFLOCK	Grefco
DICAPERL	Grefco
DIMUL	Witco
DISPERS	Tego Chemie Service
DOVERGUARD	Dover Chemical

Trade Name	Supplier
DOVERSPERSE	Dover Chemical
DOW CORNING	Dow Corning
DRIKALITE	ECC International
DUALITE	Pierce & Stevens
DURAMITE	ECC International
ECCOGARD	Eastern Color & Chemical
ECCOSTAT	Eastern Color & Chemical
ECCOWHITE	Eastern Color & Chemical
FERRO-CHAR	D.J. Enterprises
FERROSIL	Kaopolite
FILLEX	Intercorp
FIREBRAKE	U.S. Borax
FIREMASTER	Great Lakes Chemical
FIRE SHIELD	Laurel Industries
FIREX	H & C Industries
FLAME GARD	Sybron Chemicals
FLAMTARD	Alcan Chemicals
FLOMAX	J.M. Huber
FLONAC	Eckhart America
FLOW	Tego Chemie Service
FRE/HYFIL	J.M. Huber
FYARESTOR	Ferro
FYROL	Akzo Nobel Chemicals
GLIDE	Tego Chemie Service
GLYCOLUBE	Lonza
GLYCOSTAT	Lonza
GRANUSIL	Unimin
GREAT LAKES	Great Lakes Chemical
HYDRAMAX	Polymer Additives Division
HYDRAX	Polymer Additives Division
IMSIL	Unimin
IRGAFOS	Ciba Specialty Chemicals
IRGANOX	Ciba Specialty Chemicals
JETFIL	Luzenac America
KAOPOLITE	Kaopolite
KEMAMINE	Witco
KEMGARD	Sherwin-Williams
KEMOLIT	Intercorp
KOTAMITE	ECC International
KYCEROL	Rit-Chem

Trade Name	Supplier
LAROSTAT	BASF
LEA	JacksonLea
LOWINOX	Great Lakes Chemical
LSFR	Laurel Industries
LUBRIOL	Morton Plastics Additives
MAG CHEM	Martin Marietta Magnesia
MAGNIFIN	Alusuisse Aluminum
MAGNUM GLOSS	Mississippi Lime
MAGOTEX	Kaopolite
MAGSHIELD	Martin Marietta Magnesia
MARBLE DUST	ECC International
MARBLE MITE	ECC International
MARK	Witco
MARTINAL	Alusuisse Aluminum
MASTERCOLOR	Eckhart America
MASTERSAFE	Eckhart America
MAXSPERSER	Chemax
M-CURE	Sartomer
MELAPUR	DSM Melapur
MERIX	Merix Chemical
MICRAL	J.M. Huber
MICROBAN	Microban Products
MICRO-CHEK	Ferro
MICROGLASS	Fibertec
MICROPEARL	Pierce & Stevens
MICRO-WHITE	ECC International
MINEX	Unimin
MINISPHERE	Unimin
MIN-U-SIL	U.S. Silica
MISSISSIPPI	Mississippi Lime
MISTRON	Luzenac America
MIXXIM	Fairmount Chemical
MOLD PRO	Witco
NAUGARD	Uniroyal Chemical
NEVASTAIN	Neville Chemical
NOVACITE	Malvern Minerals
NOVAKUP	Malvern Minerals
NYAD	Nyco
NYCO	Nyco
NYGLOS	Nyco
OHSO	Ohio Lime
ONYX	J.M. Huber
OPACICOAT	ECC International

Trade Name	Supplier
PARABOLIX	Merix Chemical
PATIONIC	Patco Additives
PELESTAT	Tomen America
PETCAT	Laurel Industries
PIQUA	Piqua Minerals
PLASTISAN	3V Inc.
POLAQUA	ADM Tronics
POLYBOND	Uniroyal
POLYFIL	J.M. Huber
POLYGARD	Uniroyal
POLYSTAY	Goodyear
PYRO-CHEK	Ferro
PYROLUX	Sybron Chemicals
PYRONIL	Laurel Industries
RAD	Tego Chemie Service
RALOX	Raschig
ROYALTUF	Uniroyal Chemical
RYOLEX	Silbrico
SANTEL	ADM Tronics
SARET	Sartomer
SAYTEX	Albermarle
SATINTONE	Engelhard
SIL-CELL	Silbrico
SIL-CO-SIL	U.S. Silica
SILLUM	D.J. Enterprises
SILTEX	Kaopolite
SILVERBOND	Unimin
SNOWFLAKE	ECC International
SOLEM	J.M. Huber
SOLKA-FLOC	Fiber Sales & Development
SPHERIGLASS	Potters Industries
STANDART	Eckhart America
STAPA	Eckhart America
STATIC-BLOK	Amstat Industries
STATICIDE	ACL Staticide
STELLAR	Luzenac America
STONELITE	Ohio Lime
SUPERCOAT	ECC International
SUPERMITE	ECC International
SYBRON/TANATEX	Sybron Chemicals
TAMSIL	Unimin
TEGO	Tego Chemie Service
THERMOGUARD	Laurel Industries
TRANSLINK	Engelhard
TREMIN	Intercorp

Trade Name	Supplier
ULTRA FINE	Laurel Industries
ULTRANOX	GE Specialty Chemicals
UNIFOAM	Biddle Sawyer
UNISPAR	Unimin
VERTAL	Luzenac America
VULCANOL	Bayer
VULCANOX	Bayer
WESTON	GE Specialty Chemicals
WHITETEX	Engelhard
WINGSTAY	Goodyear
WOLLASTOCOAT	Nyco

Suppliers' Addresses

ADM Tronics Unlimited, Inc.
224 Pegasus Ave.
Northvale, NJ 07647
(201)-767-6040

Agrashell Inc.
5934 Keystone Dr.
Bath, PA 18014
(610)-837-6705

Akzo Nobel Chemicals Inc.
300 South Riverside Plaza
Chicago, IL 60606
(312)-906-7500/(800)-828-7929

Albright & Wilson Americas
P.O. Box 4439
Glen Allen, VA 23058
(804)-968-6300/(800)-446-3700

Alcan Chemicals
3690 Orange Place
Cleveland, OH 44122
(216)-765-2550/(800)-321-3864

AluChem Inc.
One Landy Lane
Reading, OH 45215
(513)-733-8519

American Wood Fibers
100 Anderson St.
P.O. Box 468
Schofield, WI 54476
(800)-642-5448

Amspec Chemical Corp.
751 Water St.
Gloucester City, NJ 08030
(609)-456-3930/(800)-526-7732

Amstat Industries, Inc.
3012 N. Lake Terrace
Glenview, IL 60025
(800)-783-9999

BASF Corp.
3000 Continental Drive North
Mount Olive, NJ 07828
(201)-426-2800/(800)-669-2273

Bayer Corp.
Akron, OH 44313
(330)-836-0451

Biddle Sawyer Corp.
2 Penn Plaza
New York, NY 10121
(212)-736-1580

Chartwell International Inc.
100 John Dietsch Blvd.
Attleboro Falls, MA 02763
(508)-695-1690

Chemax Inc.
P.O. Box 6067
Greenville, SC 29606
(803)-277-7000/(800)-334-6234

Ciba Specialty Chemicals
540 White Plains Rd.
Tarrytown, NY 10591
(914)-785-2000

Claremont Flock Corp.
169 Main St.
Claremont, NH 03743
(603)-542-5151

Cytec Industries, Inc.
Five Garret Mt Plaza
West Paterson, NJ 07424
(973)-357-3100/(800)-652-6013

Degussa Corp.
65 Challenger Rd.
Ridgefield Park, NJ 07660
(201)-641-6100

D.J. Enterprises Inc.
P.O. Box 31366
Cleveland, OH 44131
(216)-524-3879

Dover Chemical Corp.
3676 Davis Rd., NW
P.O. Box 40
Dover, OH 44622
(330)-343-7711/(800)-321-8805

Dow Corning Corp.
Box 0994
Midland, MI 48686
(517)-496-6000

DSM Melapur
P.O. Box 327
Addis, LA 70710
(504)-685-3000

Eastern Color & Chemical Co.
35 Livingston St.
Providence, RI 02904
(401)-331-9000

ECC International
100 Mansel Court, East
Roswell, GA 30076
(770)-594-0660

Eckhart America
72 Corwin Drive
P.O. Box 747
Painesville, OH 44077
(440)-354-0400

Endex Polymer Additives Inc.
2198 Ogden Ave.
Aurora, IL 60504
(815)-784-6286

Engelhard Corp.
101 Wood Ave.
P.O. Box 770
Iselin, NJ 08830
(800)-631-9505

Fairmount Chemical Co., Inc.
117 Blanchard St.
Newark, NJ 07105
(973)-344-5790

Ferro Corp.
7050 Krick Rd.
Walton Hills, OH 44146
(216)-641-8580

Fiber Sales & Development Corp.
P.O. Box 885
Green Brook, NJ 08812
(908)-968-5024/(800)-258-0351

Fibertec Inc.
35 Scotland Blvd.
Bridgewater, MA 02324
(508)-697-5100

Focus Chemical Corp.
B9 Orchard Park
875 Greenland Road
Portsmouth, NH 03801
(603)-430-9802

Franklin Industrial Minerals
612 Tenth Ave. North
Nashville, TN 37203
(615)-259-4222/(800)-626-8147

GE Silicones
260 Hudson River Rd.
Waterford, NY 12188
(518)-237-3330/(800)-255-8886

GE Specialty Chemicals
Avery St.
Parkersburg, WV 26102
(304)-424-5411/(800)-872-0022

Goodyear Chemicals
1452 E. Archwood Ave.
Akron, OH 44316
(330)-796-7110

Great Lakes Chemical Corp.
P.O. Box 2200
West Lafayette, IN 47906
(765)-497-6100/(800)-428-7947

Grefco Minerals, Inc.
333 E. Hwy 246
Lompoc, CA 93436
(310)-577-0700

H & C Industries, Inc.
1311 Crenshaw Blvd.
Torrance, CA 90501
(310)-328-3683

J.M. Huber Corp.
P.O. Box 310
Revolution St.
Havre de Grace, MD 21078
(410)-939-3500

Intercorp Inc.
3628 West Pierce St.
Milwaukee, WI 53215
(414)-383-2020

JacksonLea
75 Progress Lane
P.O. Box 71
Waterbury, CT 06720
(203)-753-5116

Kaopolite Inc.
100 Mansell Court East
Roswell, GA 30076
(888)-567-6541

Laurel Industries Inc.
30195 Chagrin Blvd.
Cleveland, OH 44124
(216)-831-5747/(800)-221-1304

Lonza Inc.
17-17 Route 208
Fair Lawn, NJ 07410
(201)-794-2780/(800)-777-1875

Luzenac America Inc.
9000 E. Nichols Ave.
Englewood, CO 80112
(303)-643-0040

Malvern Minerals Co.
P.O. Drawer 1238
Hot Springs, AR 71902
(501)-623-8893

Martin Marietta Magnesia Spec.
P.O. Box 15470
Baltimore, MD 21220
(410)-780-5500/(800)-648-7400

Mayzo Inc.
6577 Peachtree Industrial Blvd.
Norcross, GA 30092
(770)-449-9066

Merix Chemical Co.
2234 E 75 St.
Chicago, IL 60649
(773)-221-8242

Mississippi Lime Co.
7 Alby St.
Alton, IL 62002
(618)-465-7741/(800)-437-5463

Morton Plastics Additives
2000 West St.
Cincinnati, OH 45215
(513)-733-2100

Neville Chemical Co.
2800 Neville Rd.
Pittsburgh, PA 15225
(412)-331-4200

Nyacol Products Inc.
Megunco Rd.
Ashland, MA 01721
(508)-881-2220

NYCO Minerals Inc.
P.O. Box 368
124 Mountain View Drive
Willsboro, NY 12996
(518)-963-4262

Ohio Lime Inc.
P.O. Box 708
Bettsville, OH 44815
(419)-849-2321/(800)-445-3930

Patco Additives Division
3947 Broadway
Kansas City, MO 64111
(816)-561-9050/(800)-669-2250

Pierce & Stevens Corp.
710 Ohio St.
P.O. Box 1092
Buffalo, NY 14240
(716)-856-4910/(800)-888-4910

Piqua Minerals
1750 W Statler Road
Piqua, OH 45356
(800)-338-2962

Potters Industries Inc.
Affiliate of PQ Corp.
P.O. Box 840
Valley Forge, PA 19482
(610)-651-4200/(800)-944-7411

Raschig Corp.
26032 Detroit Rd.
Westlake, OH 44145
(440)-356-7502

Rit-Chem Co. Inc.
109 Wheeler Ave.
P.O. Box 435
Pleasantville, NY 10570
(914)-769-9110

Sangi America Inc.
516 N Pennfield Place
Thousand Oaks, CA 91360
(805)-379-8590

Sartomer Co. Inc.
Oaklands Corporate Center
502 Thomas Jones Way
Exton, PA 19341
(610)-363-4100/(800)-345-8247

Shamokin Filler Co. Inc.
Venn Access Rd.
Box 568
Shamokin, PA 17872
(717)-644-0437

Sherwin Williams Co.
5501 Cass Ave.
Cleveland, OH 44102
(800)-524-5979

Silbrico Corp.
6300 River Rd.
Hodgkins, IL 60525
(708)-354-3350/(800)-323-4287

Sybron Chemicals Inc.
Hwy 29
P.O. Box 125
Wellford, SC 29385
(864)-439-6333/(800)-677-3500

3V Inc.
1500 Harbor Blvd.
Weehawken, NJ 07087
(201)-865-3600/(800)-441-5156

Tomah Products Inc.
1012 Terra Drive
Milton, WI 53563
(608)-868-6811/(800)-441-0708

Unimin Specialty Minerals Inc.
Rt. 127
P.O. Box 33
Elco, IL 62929
(800)-743-1519

Uniroyal Chemical Co. Inc.
Benson Rd.
Middlebury, CT 06749
(203)-573-2000/(800)-243-3024

U.S. Borax Inc.
26877 Tourney Rd.
Valencia, CA 91355
(805)-287-5400

U.S. Silica Co.
P.O. Box 187
Berkeley Springs, WV 25411
(304)-258-2500/(800)-243-7500

Whittemore Co. Inc.
30 Glenn St.
Lawrence, MA 01843
(508)-681-8833

Witco Corp.
One American Lane
Greenwich, CT 06831
(800)-494-8737